Learning Real-time Processing with Spark Streaming

Building scalable and fault-tolerant streaming applications made easy with Spark Streaming

Sumit Gupta

BIRMINGHAM - MUMBAI

Learning Real-time Processing with Spark Streaming

Copyright © 2015 Packt Publishing

All rights reserved. No part of this book may be reproduced, stored in a retrieval system, or transmitted in any form or by any means, without the prior written permission of the publisher, except in the case of brief quotations embedded in critical articles or reviews.

Every effort has been made in the preparation of this book to ensure the accuracy of the information presented. However, the information contained in this book is sold without warranty, either express or implied. Neither the author, nor Packt Publishing, and its dealers and distributors will be held liable for any damages caused or alleged to be caused directly or indirectly by this book.

Packt Publishing has endeavored to provide trademark information about all of the companies and products mentioned in this book by the appropriate use of capitals. However, Packt Publishing cannot guarantee the accuracy of this information.

First published: September 2015

Production reference: 1230915

Published by Packt Publishing Ltd.
Livery Place
35 Livery Street
Birmingham B3 2PB, UK.

ISBN 978-1-78398-766-5

www.packtpub.com

Credits

Author
Sumit Gupta

Reviewers
David Morales De Frías
Sujit Pal
Antonio Navarro Pérez

Commissioning Editor
Kunal Parikh

Acquisition Editor
Larissa Pinto

Content Development Editor
Rashmi Suvarna

Technical Editor
Siddhi Rane

Copy Editors
Kevin McGowan
Swati Priya

Project Coordinator
Judie Jose

Proofreader
Safis Editing

Indexer
Priya Sane

Graphics
Disha Haria
Jason Monteiro
Abhinash Sahu

Production Coordinator
Shantanu N. Zagade

Cover Work
Shantanu N. Zagade

About the Author

Sumit Gupta is a seasoned professional, innovator, and technology evangelist with over 100 months of experience in architecting, managing, and delivering enterprise solutions revolving around a variety of business domains such as hospitality, healthcare, risk management, insurance, and so on. He is passionate about technology with an overall 14 years of hands-on experience in the software industry and has been using big data and cloud technologies over the past 4 to 5 years to solve complex business problems.

He is also the author of *Neo4j Essentials* and *Building Web Applications with Python and Neo4j* both by Packt Publishing.

> I want to acknowledge and express my gratitude to everyone who supported me in authoring this book. I am thankful for their aspiring guidance, and for their valuable, constructive, and friendly advice.

About the Reviewers

David Morales De Frías is the lead big data architect at Stratio, the creators of a pure Apache Spark big data platform. He has led the development of open source projects based on Spark Streaming, such as Stratio Sparkta for real-time aggregation, and Stratio Streaming for complex event processing, being identified as a thought leader by the Spark Streaming community. He has also been involved in the design, development, and deployment of real-time big data powered solutions from customer intelligence to fraud detection at global enterprises, including banks, telcos, commerce, and others.

Sujit Pal works at Elsevier Labs, a research and development group within the Reed-Elsevier PLC (RELX) Group. His interests are in the fields of information retrieval, distributed processing, ontology development, natural language processing, and machine learning. Sujit uses Spark with Scala and Python to analyze and extract intelligence from large volumes of STM (science, technology, and medicine) content and user-generated metadata. He believes in lifelong learning and blogs about his experiences at http://sujitpal.blogspot.in/.

Antonio Navarro Pérez has been working in a huge variety of programming projects, from management applications, REST APIs, mobile payments processing to big data projects and NewSQL databases. He has been working very closely with Spark Streaming, complex event processing, real-time data processing, and more widely expanding the Spark SQL functionality. He is a passionate software developer and open source advocate.

www.PacktPub.com

Support files, eBooks, discount offers, and more

For support files and downloads related to your book, please visit www.PacktPub.com.

Did you know that Packt offers eBook versions of every book published, with PDF and ePub files available? You can upgrade to the eBook version at www.PacktPub.com and as a print book customer, you are entitled to a discount on the eBook copy. Get in touch with us at service@packtpub.com for more details.

At www.PacktPub.com, you can also read a collection of free technical articles, sign up for a range of free newsletters and receive exclusive discounts and offers on Packt books and eBooks.

https://www2.packtpub.com/books/subscription/packtlib

Do you need instant solutions to your IT questions? PacktLib is Packt's online digital book library. Here, you can search, access, and read Packt's entire library of books.

Why subscribe?

- Fully searchable across every book published by Packt
- Copy and paste, print, and bookmark content
- On demand and accessible via a web browser

Free access for Packt account holders

If you have an account with Packt at www.PacktPub.com, you can use this to access PacktLib today and view 9 entirely free books. Simply use your login credentials for immediate access.

Table of Contents

Preface	**v**
Chapter 1: Installing and Configuring Spark and Spark Streaming	**1**
Installation of Spark	**2**
Hardware requirements	2
CPU	3
RAM	3
Disk	4
Network	4
Operating system	4
Software requirements	5
Spark	5
Java	6
Scala	6
Eclipse	7
Installing Spark extensions – Spark Streaming	8
Configuring and running the Spark cluster	**8**
Your first Spark program	**12**
Coding Spark jobs in Scala	12
Coding Spark jobs in Java	16
Tools and utilities for administrators/developers	**18**
Cluster management	19
Submitting Spark jobs	20
Troubleshooting	**20**
Configuring port numbers	21
Classpath issues – class not found exception	21
Other common exceptions	21
Summary	**22**

Table of Contents

Chapter 2: Architecture and Components of Spark and Spark Streaming — 23
Batch versus real-time data processing — 24
- Batch processing — 25
- Real-time data processing — 26

Architecture of Spark — 28
- Spark versus Hadoop — 29
- Layered architecture – Spark — 30

Architecture of Spark Streaming — 32
- What is Spark Streaming? — 32
- High-level architecture – Spark Streaming — 32

Your first Spark Streaming program — 35
- Coding Spark Streaming jobs in Scala — 35
- Coding Spark Streaming jobs in Java — 38
- The client application — 40
- Packaging and deploying a Spark Streaming job — 42

Summary — 44

Chapter 3: Processing Distributed Log Files in Real Time — 45
Spark packaging structure and client APIs — 46
- Spark Core — 48
  - SparkContext and Spark Config – Scala APIs — 48
  - SparkContext and Spark Config – Java APIs — 49
  - RDD – Scala APIs — 49
  - RDD – Java APIs — 50
  - Other Spark Core packages — 51
- Spark libraries and extensions — 54
  - Spark Streaming — 54
  - Spark MLlib — 55
  - Spark SQL — 56
  - Spark GraphX — 57

Resilient distributed datasets and discretized streams — 58
- Resilient distributed datasets — 58
  - Motivation behind RDD — 58
  - Fault tolerance — 59
  - Transformations and actions — 60
  - RDD storage — 61
  - RDD persistence — 61
  - Shuffling in RDD — 62
- Discretized streams — 63

Data loading from distributed and varied sources — 65
- Flume architecture — 67
- Installing and configuring Flume — 69
- Configuring Spark to consume Flume events — 73

Packaging and deploying a Spark Streaming job	76
Overall architecture of distributed log file processing	77
Summary	**78**

Chapter 4: Applying Transformations to Streaming Data — 79

Understanding and applying transformation functions	**80**
Simulating log streaming	80
Functional operations	82
Transform operations	89
Windowing operations	90
Performance tuning	**93**
Partitioning and parallelism	93
Serialization	94
Spark memory tuning	95
Garbage collection	95
Object sizes	95
Executor memory and caching RDDs	96
Summary	**97**

Chapter 5: Persisting Log Analysis Data — 99

Output operations in Spark Streaming	**100**
Integration with Cassandra	**110**
Installing and configuring Apache Cassandra	110
Configuring Spark for integration with Cassandra	112
Coding Spark jobs for persisting streaming web logs in Cassandra	113
Summary	**120**

Chapter 6: Integration with Advanced Spark Libraries — 121

Querying streaming data in real time	**122**
Understanding Spark SQL	123
Integrating Spark SQL with streams	129
Graph analysis – Spark GraphX	**135**
Introduction to the GraphX API	137
Integration with Spark Streaming	140
Summary	**146**

Chapter 7: Deploying in Production — 147

Spark deployment models	**148**
Deploying on Apache Mesos	149
Installing and configuring Apache Mesos	150
Integrating and executing Spark applications on Apache Mesos	152
Deploying on Hadoop or YARN	154
High availability and fault tolerance	**158**
High availability in the standalone mode	158

>   High availability in Mesos or YARN 160
>   Fault tolerance 161
>   Fault tolerance in Spark Streaming 162
>   **Monitoring streaming jobs** **165**
>   Application or job UI 166
>   Integration with other monitoring tools 170
>   **Summary** **171**
> **Index** **173**

Preface

Processing large data and producing business insights is one of the popular use cases, performed for deriving business intelligence (BI) over the historical data. Enterprises have been focusing on developing data warehouses (https://en.wikipedia.org/wiki/Data_warehouse) where they want to store the data fetched from every possible data source and leverage various BI tools for providing analytics/analysis over the data stored in these data warehouses. But developing data warehouses is a complex, time-consuming, and costly process. It may take months or sometimes years too.

No doubt the emergence of Hadoop and its ecosystem has provided a new architecture for solving large data problems. It provides a low cost and scalable solution, which may process terabytes of data in few hours, which earlier could have taken days.

The following illustration shows the typical Apache Hadoop ecosystem and its various components used to develop solutions for large data problems:

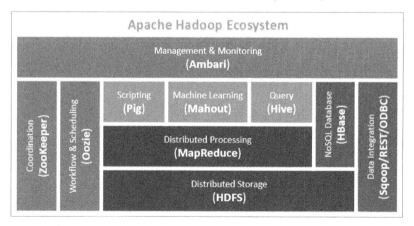

Image credits: http://www.dezyre.com/article/big-data-and-hadoop-training-hadoop-components-and-architecture/114

Preface

This was only one side of the coin where Hadoop was meant for batch processes, while there were other business use cases, which required to produce business insights in real or near real time (sub seconds SLA) too. This was called as real-time business intelligence (RTBI) or near real-time business intelligence (NRTBI), `https://en.wikipedia.org/wiki/Real-time_business_intelligence`. It was also termed as "fast data", where it implies the ability to make near real-time decisions and enable orders of magnitude improvements in elapsed time to decisions for businesses.

A number of powerful, easy-to-use open source platforms have emerged to solve these enterprise real-time data use cases. Two of the most notable ones are Apache Storm and Apache Spark, which offer real-time processing capabilities to a much wider range of potential users. Both projects are part of the Apache Software Foundation, and while the two tools provide overlapping capabilities, each of them have distinctive features and roles to play.

Apache Storm is an excellent framework for reliable distributed stream processing or it would be appropriate to say that it is suitable for CEP style processing (`https://en.wikipedia.org/wiki/Complex_event_processing`) with its own deployment process. It worked for majority of the near/real-time use cases, but it failed to provide answers to the questions like:

- How if the same data needs to be processed in batch and near real-time? Does it need deployment of two different frameworks (Hadoop and Storm)?
- How about merging of streams coming from two different data sources?
- Apart from Java, can I use another programming language?
- Can we integrate near real-time streams with other systems such as graphs, SQL, Hive, and so on?
- How about near real-time recommendations, clustering, or classifications?

Apache Spark was the answer to all the preceding questions. It not only retained the benefits of Hadoop and Storm, but at the same time, it provided a unified framework where you can write your code in a variety of programming languages such as Python, Java, or Scala and reuse the same piece of code across the streaming and batch use cases. It also provided various libraries and extensions like these:

- Spark GraphX: For developing graphs
- Spark DataFrames and SQL: Executing SQL queries
- Spark MLlib: Executing machine learning algorithms for recommendations, clustering and classifications
- Spark Streaming: For handling streaming data in near real time

One of the notable features of Apache Spark was the interoperability of all these libraries and extensions. For example, the data received from near real-time streams can be converted into graphs or may be analyzed using SQL, or we may execute machine learning algorithms for providing recommendation, clustering, or classifications.

Interesting, isn't it?

Apache Spark started out as a project of AMPLab at the University of California at Berkeley before joining the Apache Incubator and ultimately graduating as a top-level project in February 2014. Spark is more of a general-purpose distributed computing platform, which supports both batch as well as near real-time data processing.

Let's move forward and jump into the nitty-gritties of real-time processing with Spark Streaming.

In subsequent chapters, we will cover the various aspects dealing with installation and configuration of Spark Streaming, its architecture, Spark Streaming operations, integration with other Spark libraries, and NoSQL databases, and finally, the deployment aspects of Spark Streaming in the production environment.

Preface

What this book covers

Chapter 1, *Installing and Configuring Spark and Spark Streaming*, details the installation process and configuration of Spark and Spark Streaming. It talks about the prerequisites required for running Spark jobs. It also details the various tools and utilities packaged with Spark along with their usage and function. This chapter also introduces and helps developers to write their first Spark job in Java and Scala and execute it on the cluster. Finally, it ends with troubleshooting tips and tricks for most common and frequent errors encountered with Spark and Spark Streaming.

Chapter 2, *Architecture and Components of Spark and Spark Streaming*, starts with the introduction about the complexities and overall paradigm of batch and real-time data processing. Then, it also elaborates the architecture of Spark, where Spark Streaming fits into the overall architecture. Finally, it helps developers to write and execute their Spark Streaming programs on the Spark cluster.

Chapter 3, *Processing Distributed Log Files in Real Time*, discusses the packaging structure and various important APIs of Spark and Spark Streaming. Then, it further discusses about the two core components of Spark and Spark Streaming—resilient distributed datasets and discretized streams. Further, it introduces distributed log processing use case, where we develop and execute Spark Streaming jobs to load the data from distributed data sources with the help of Apache Flume.

Chapter 4, *Applying Transformations to Streaming Data*, discusses the various transformation operations exposed by Spark Streaming—functional, transform, and windowing operations, and then further enhances our distributed log processing use case by applying all those transformation operations on the streaming data. It also discusses the various factors and considerations for improving performance of our Spark Streaming jobs.

Chapter 5, *Persisting Log Analysis Data*, talks about the output operations exposed by Spark Streaming. It shows the integration with Apache Cassandra for persisting the distributed log data received and processed by our Spark Streaming job.

Chapter 6, *Integration with Advanced Spark Libraries*, enhances distributed log file processing use cases and discusses the integration of Spark Streaming with advance Spark libraries such as GraphX and Spark SQL.

Chapter 7, *Deploying in Production*, discusses the various aspects that need to be considered while deploying Spark Streaming application in production such as high availability, fault tolerance, monitoring, and so on. It also discusses the process for deploying Spark Streaming jobs in other cluster computing frameworks such as YARN and Apache Mesos.

What you need for this book

You should have programming experience in Scala and some basic knowledge and understanding of any distributed computing platform such as Apache Hadoop.

Who this book is for

This book is aimed at competent developers who have good knowledge and understanding of Scala to allow efficient programming of core elements and applications.

If you are reading this book, then you probably already have sufficient knowledge of Scala. This book will cover real-time data processing with the help of Spark Streaming. It will also discuss the distributed log stream processing in near real-time with the help of various APIs and operations exposed by Spark Streaming.

Conventions

In this book, you will find a number of text styles that distinguish between different kinds of information. Here are some examples of these styles and an explanation of their meaning.

Code words in text, database table names, folder names, filenames, file extensions, pathnames, dummy URLs, user input, and Twitter handles are shown as follows: "Increase the `ulimit` on your Linux OS by executing `sudo ulimit -n 20000`."

A block of code is set as follows:

```
public class JavaFirstSparkExample {

  public static void main(String args[]){
    //Java Main Method
  }
}
```

Any command-line input or output is written as follows:

```
export FLUME_HOME=<path of extracted Flume binaries>
```

New terms and **important words** are shown in bold. Words that you see on the screen, for example, in menus or dialog boxes, appear in the text like this: "Clicking the **Next** button moves you to the next screen."

> Warnings or important notes appear in a box like this.

> Tips and tricks appear like this.

Reader feedback

Feedback from our readers is always welcome. Let us know what you think about this book—what you liked or disliked. Reader feedback is important for us as it helps us develop titles that you will really get the most out of.

To send us general feedback, simply e-mail feedback@packtpub.com, and mention the book's title in the subject of your message.

If there is a topic that you have expertise in and you are interested in either writing or contributing to a book, see our author guide at www.packtpub.com/authors.

Customer support

Now that you are the proud owner of a Packt book, we have a number of things to help you to get the most from your purchase.

Downloading the example code

You can download the example code files from your account at http://www.packtpub.com for all the Packt Publishing books you have purchased. If you purchased this book elsewhere, you can visit http://www.packtpub.com/support and register to have the files e-mailed directly to you.

Errata

Although we have taken every care to ensure the accuracy of our content, mistakes do happen. If you find a mistake in one of our books—maybe a mistake in the text or the code—we would be grateful if you could report this to us. By doing so, you can save other readers from frustration and help us improve subsequent versions of this book. If you find any errata, please report them by visiting http://www.packtpub.com/submit-errata, selecting your book, clicking on the **Errata Submission Form** link, and entering the details of your errata. Once your errata are verified, your submission will be accepted and the errata will be uploaded to our website or added to any list of existing errata under the Errata section of that title.

To view the previously submitted errata, go to https://www.packtpub.com/books/content/support and enter the name of the book in the search field. The required information will appear under the **Errata** section.

Piracy

Piracy of copyrighted material on the Internet is an ongoing problem across all media. At Packt, we take the protection of our copyright and licenses very seriously. If you come across any illegal copies of our works in any form on the Internet, please provide us with the location address or website name immediately so that we can pursue a remedy.

Please contact us at copyright@packtpub.com with a link to the suspected pirated material.

We appreciate your help in protecting our authors and our ability to bring you valuable content.

Questions

If you have a problem with any aspect of this book, you can contact us at questions@packtpub.com, and we will do our best to address the problem.

1
Installing and Configuring Spark and Spark Streaming

Apache Spark (http://spark.apache.org/) is a general-purpose, open source cluster computing framework developed in the AMPLab in UC Berkeley in 2009.

The emergence of Spark has not only opened new data processing possibilities for a variety of business use cases but at the same time introduced a unified platform for performing various batch and real-time operations using a common framework. Depending on the user and business needs, data can be consumed or processed every second (even less) or maybe every day, which is in harmony with the needs of enterprises.

Spark, being a general-purpose distributed processing framework, enables **Rapid Application Development** (**RAD**) and at the same time it also allows the reusability of code across batch and streaming applications. One of the most enticing features of Spark is that you can code on your desktop or laptop and it can also be deployed on top of several other cluster managers provided by Apache Mesos (http://mesos.apache.org/) or Apache Hadoop YARN (https://hadoop.apache.org/) without any changes.

We will talk more about Spark and its features in the subsequent chapters but let's move ahead and prepare (install and configure) our environment for development on Apache Spark.

This chapter will help you understand the paradigm, applicability, aspects and characteristics of Apache Spark and Spark Streaming. It will also guide you through the installation process and running your first program using the Spark framework. At the end of this chapter, your work environment will be fully functional and ready to explore and develop applications using the Spark framework. This chapter will cover the following points:

- Installation of Spark
- Configuring and running a Spark cluster
- Your first Spark program
- Tools and utilities for administrators
- Troubleshooting

Installation of Spark

In this section we will discuss the various aspects of Spark installation and its dependent components.

Spark supports a variety of hardware and software platforms. It can be deployed on commodity hardware and also supports deployments on high-end servers. Spark clusters can be provisioned either on cloud or on-premises. Though there is no single configuration or standards which can guide us through the requirement of Spark but still we can define "must to have" versus "good to have" and the rest varies on the requirements imposed by the use cases.

We will discuss deployment aspects more in *Chapter 7, Deploying in Production*, but let's move forward and understand the hardware and software requirements for developing applications on Spark.

Hardware requirements

In this section we will discuss the hardware required for batch and real-time applications developed on Spark.

CPU

Spark provides data processing in batch and real-time and both kinds of workloads are CPU-intensive. In large scale deployments, there has to be perfect management and utilization of computing resources. Spark solves this challenge by reducing the sharing or context switching between the threads. The objective is to provide sufficient computing resources to each thread, so that it can run independently and produce results in a timely manner. The following are the recommended requirements for CPU for each machine which will be part of a Spark cluster:

- Must have: Dual core (2 cores)
- Good to have: 16 cores

RAM

Real-time data processing or low latency jobs mandate that all reads/writes happen from memory itself. Any reads/writes happening from disk may impact performance.

Spark provides the optimal performance for memory intensive jobs by caching datasets within the memory itself, so that the data can be directly read or processed from memory itself and there are very few or no reads from disks.

The general rule for memory is "the more, the better" but it depends on your use case and application. Spark is implemented in Scala (http://www.scala-lang.org/), which requires JVM as a runtime environment for deploying Spark and, as is true for other Java-based applications, the same is applicable for Spark. We need to provide optimum memory for optimal performance. As a general rule, we should allocate only 75 percent of available memory to our Spark application and the rest should be left for the OS and other system processes. Considering all aspects and constraints exposed by the Java garbage collector, the following are the memory requirements:

- Must have: 8 GB
- Good to have: 24 GB

Disk

Everything cannot be fitted into memory and eventually you need a persistent storage area (disk) for storing the data which cannot be fitted into the memory. Spark automatically spills the datasets that do not fit in memory either to the disk or re-computes on the fly when needed. Again the exact size of disks depends on the data size of your application but we would recommend the following specifications:

- Must have: SATA drives with 15k RPM, with minimum capacity of 1-2 TBs each.
- Good to have: Non-RAID architecture — deploying **Just Bunch of Disks (JBOD)** without any data redundancy capabilities. SSDs are preferred for higher throughput and better response times.

> Refer to the following link for non-RAID architectures:
> http://en.wikipedia.org/wiki/Non-RAID_drive_architectures

Network

Movement of datasets from one node to another is an intrinsic feature of any distributed computing framework. If your network is slow then it will eventually impact the performance of your job.

Spark provides a "scale out" architecture for when an application runs of multiple nodes and is network-bound for large computations which span over multiple machines. Here are the recommended specifications for the network bandwidth allocated between the nodes of a Spark cluster.

- Must have: 1 Gbps
- Good to have: 10 Gbps

Operating system

Spark follows the principle of code once and deploy anywhere. Since Spark is coded in Scala, Spark jobs can be deployed on a large number of operating systems. The following are the various flavors of OS recommended for Spark deployment:

- Production: Linux, HP UX
- Development: Windows XP/7/8, Mac OS X, Linux

In this section we have discussed the hardware prerequisites for setting up the Spark cluster. Let's move forward and discuss the software requirements for developing, building and deploying our Spark applications.

Software requirements

In this section we will talk about the software required for developing and deploying Spark-based applications.

Spark Core is coded in Scala but it offers several development APIs in different languages such as Scala, Java and Python, so that you can choose your preferred weapon for coding. The dependent software may vary based on the programming languages but still there are common sets of software for configuring the Spark cluster and then language-specific software for developing Spark jobs in specific programming languages. Spark also supports deployment and development on Windows and Linux but, for brevity, we will discuss the installation steps only for the Linux-based operating systems specifically for Java and Scala.

Let's install all the required software which we need for developing Spark-based applications in Scala and Java.

Spark

Perform the following steps to install Spark:

1. Download Spark compressed tarball from http://d3kbcqa49mib13.cloudfront.net/spark-1.3.0-bin-hadoop2.4.tgz.
2. Create a new directory Spark-1.3.0 on your local file system and extract Spark tarball into this directory.
3. Execute the following command on your Linux shell for setting SPARK_HOME as an environment variable:

   `export SPARK_HOME=<Path of Spark install Dir>`

Installing and Configuring Spark and Spark Streaming

4. Now browse your directory `SPARK_HOME` and it should be similar to the following illustration:

```
sumit@local : ls -ltr
total 360
drwxrwxr-x. 2 impadmin impadmin   4096 Mar  6 06:01 sbin
-rw-rw-r--. 1 impadmin impadmin    134 Mar  6 06:01 RELEASE
-rw-rw-r--. 1 impadmin impadmin   3629 Mar  6 06:01 README.md
drwxrwxr-x. 7 impadmin impadmin   4096 Mar  6 06:01 python
-rw-rw-r--. 1 impadmin impadmin  46083 Mar  6 06:01 LICENSE
drwxrwxr-x. 3 impadmin impadmin   4096 Mar  6 06:01 ec2
drwxrwxr-x. 3 impadmin impadmin   4096 Mar  6 06:01 data
drwxrwxr-x. 2 impadmin impadmin   4096 Mar  6 06:01 conf
-rw-rw-r--. 1 impadmin impadmin 250524 Mar  6 06:01 CHANGES.txt
drwxrwxr-x. 2 impadmin impadmin   4096 Mar  6 06:01 bin
-rw-rw-r--. 1 impadmin impadmin  22559 Mar  6 06:01 NOTICE
drwxrwxr-x. 2 impadmin impadmin   4096 Mar  6 06:01 lib
drwxrwxr-x. 3 impadmin impadmin   4096 Mar  6 06:01 examples
sumit@local :
```

Java

Perform the following steps for installing Java:

1. Download and install Oracle Java 7 from `http://www.oracle.com/technetwork/java/javase/install-linux-self-extracting-138783.html`.

2. Execute the following command on your Linux shell for setting `JAVA_HOME` as an environment variable:

   `export JAVA_HOME=<Path of Java install Dir>`

Scala

Perform the following steps for installing Scala:

1. Download Scala 2.10.5 compressed tarball from `http://downloads.typesafe.com/scala/2.10.5/scala-2.10.5.tgz?_ga=1.7758962.1104547853.1428884173`.

2. Create a new directory `Scala-2.10.5` on your local filesystem and extract Scala tarball into this directory.

3. Execute the following commands on your Linux shell for setting `SCALA_HOME` as an environment variable and add the Scala compiler in the system `$PATH`.

   `export SCALA_HOME=<Path of Scala install Dir>`

   `export PATH = $PATH:$SCALA_HOME/bin`

Chapter 1

4. Next, execute the following command to ensure that Scala runtime and Scala compiler is available and version is 2.10.x:

```
sumit@local : scala -version
Scala code runner version 2.10.5 -- Copyright 2002-2013, LAMP/EPFL
sumit@local : scalac -version
Scala compiler version 2.10.5 -- Copyright 2002-2013, LAMP/EPFL
sumit@local :
```

> Spark 1.3.0 is packaged and supports the 2.10.5 version of Scala, so it is advisable to use the same version to avoid any runtime exceptions due to mismatch of libraries.

Eclipse

Perform the following steps to install Eclipse:

1. Based on your hardware configuration, download Eclipse Luna (4.4) from http://www.eclipse.org/downloads/packages/eclipse-ide-java-ee-developers/lunasr2:

2. Next, install IDE for Scala in Eclipse itself, so that we can write and compile our Scala code inside Eclipse (http://scala-ide.org/download/current.html).

And we are done with the installation of all the required software!

[7]

Installing Spark extensions – Spark Streaming

Core Spark packages provide the functionality of distributed processing of datasets in batches. It is often referred to as **batch processing**. At the same time Spark also provides extensions like MLlib, GraphX, and so on for other desired functionalities.

Spark Streaming is one such extension for processing and streaming data and it is packaged with Spark itself. We do not have to install anything separate to install Spark Streaming. Spark Streaming is an API which is packaged and integrated with Spark itself. In subsequent chapters we will discuss the Spark Streaming API and its usages more.

In the previous section we have installed all the required software. Let's move forward and configure our Spark cluster for the execution of Spark jobs.

Configuring and running the Spark cluster

In this section, we will configure our Spark cluster so that we can deploy and execute our Spark application.

Spark essentially enables the distributed execution of a given piece of code. Though we will talk about Spark architecture in the next chapter, let's briefly talk about the major components which need to be configured for setting up the Spark cluster.

The following are the high-level components involved in setting up the Spark cluster:

- **Driver**: It is the client program which defines `SparkContext`. It connects to the cluster manager and requests resources for further execution of the jobs in distributed mode.

- **Cluster manager / Spark master**: Cluster manager manages and allocates the required system resources to the Spark jobs. Furthermore, it coordinates and keeps track of the live/dead nodes in a cluster. It enables the execution of jobs submitted by the driver on worker nodes (also called Spark workers) and finally tracks and shows the status of various jobs running by the worker nodes.

- **Spark worker**: Worker actually executes the business logic submitted by the driver. Spark workers are abstracted from the Spark driver and are allocated to the driver by the cluster manager dynamically.

The following diagram shows the high-level components of Spark and the way they work in combination for the execution of the submitted jobs:

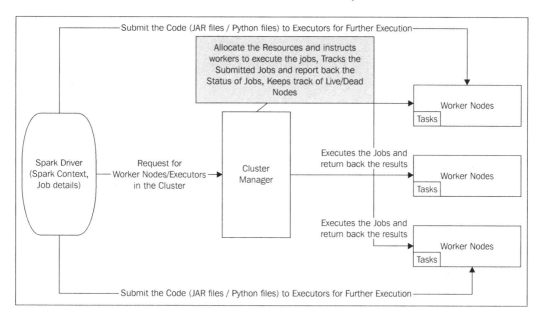

Now we know about the different components of the Spark cluster, let's move forward and set up these different components and bring up the Spark cluster.

Spark supports three different deployment models for configuring the cluster and different components of Spark in production and other environments:

- **Standalone mode**: The core Spark distribution contains the required APIs to create an independent, distributed and fault-tolerant cluster without any external or third-party libraries or dependencies

> Standalone mode should not confused with local mode. In local mode Spark jobs can be executed on a local machine without any special cluster setup, just passing `local[N]` as the master URL, where N is the number of parallel threads.

Installing and Configuring Spark and Spark Streaming

- **Apache Mesos** (http://mesos.apache.org/): This is a distributed general computing framework which abstracts out system resources like CPU and memory and enables the distributed execution of the submitted jobs
- **Hadoop YARN** (http://hadoop.apache.org/docs/current/hadoop-yarn/hadoop-yarn-site/YARN.html): Spark can also utilize the resource manager of Hadoop-2 for acquiring the cluster resources and scheduling the Spark jobs

We will discuss the deployment models in detail in *Chapter 7, Deploying in Production*, but here we will discuss the bare minimum configuration and the steps for configuring our Spark cluster using **standalone mode**, so that we can quickly move forwards towards the next section in which we will write and execute our first Spark program.

Perform the following steps to bring up an independent cluster using Spark binaries:

1. The first step in setting up the Spark cluster is to bring up the master node which will track and allocate the systems resource. Open your Linux shell and execute the following command:

   ```
   $SPARK_HOME/sbin/start-master.sh
   ```

   The preceding command will bring up your master node and it will also enable a UI—Spark UI for monitoring the nodes/jobs in Spark cluster— http://<host>:8080/. "<host>" is the domain name of the machine on which the master is running.

2. Next, let's bring up our worker nodes, which will execute our Spark jobs. Execute the following command on the same Linux shell:

   ```
   $SPARK_HOME/bin/spark-class org.apache.spark.deploy.worker.Worker <Spark-Master> &
   ```

3. In the preceding command, replace the `<Spark-Master>` with the Spark URL shown at the top of the Spark UI, just besides Spark master at. The preceding command will start the Spark worker process in the background and the same will also be reported in the Spark UI.

Chapter 1

The Spark UI shown in the preceding illustration reports the following statuses in three different sections:

- **Workers**: Reports the health of a worker node—alive or dead and also provides drill-down to query the status and details logs of the various jobs executed by that specific worker node
- **Running Applications**: Shows the applications which are currently being executed in the cluster and also provides drill-down and enables viewing of application logs
- **Completed Application**: Same functionality as **Running Applications,** the only difference being that it shows the jobs which are finished

And we are done!

Our Spark cluster is up and running and ready to execute our Spark jobs with one worker node.

Let's move forward and write our first Spark application in Scala and Java and further execute it on our newly created cluster.

> **Downloading the example code**
>
> You can download the example code files from your account at http://www.packtpub.com for all the Packt Publishing books you have purchased. If you purchased this book elsewhere, you can visit http://www.packtpub.com/support and register to have the files e-mailed directly to you.

Your first Spark program

In this section we will discuss the basic terminology used in Spark and then we will code and deploy our first Spark application using Scala and Java.

Now as we have configured our Spark cluster, we are ready to code and deploy our Spark jobs but, before moving forward, let's talk about a few important components of Spark:

- **RDD**: Spark works on the concept of **RDD (Resilient Distributed Datasets)**. All data which needs to be processed in Spark needs to be converted into RDD and then it is loaded into the Spark cluster for further processing. RDD is a distributed memory abstraction that lets programmers perform in-memory computations on large clusters in a fault-tolerant manner. Spark provides various ways to create RDDs such as RDDs using Hadoop input formats. Raw text or binary files can also be converted into RDDs.

   We will talk more about RDDs in *Chapter 3, Processing Distributed Log Files in Real Time*, but the preceding description should be sufficient to understand subsequent examples.

- **SparkContext**: SparkContext is the key to access all features exposed by the Spark framework. It is the main entry point for any application for creating connections to a Spark cluster, access methods for creating RDDs and so on. The only constraint with SparkContext is that there cannot be more than one SparkContext in a given JVM but multiple contexts can be created in different JVMs. Eventually this constraint may be removed in future releases of Spark (`https://issues.apache.org/jira/browse/SPARK-2243`).

Coding Spark jobs in Scala

In this section we will code our Spark jobs in Scala. This will be our first Spark job so we will keep it simple and count the number of lines in a given text file.

Perform the following steps to code the Spark example in Scala which counts the number of lines given in a text file:

1. Open Eclipse and create a Scala project called `Spark-Examples`.
2. Expand your newly created project and modify the version of **Scala library container** to 2.10. This is done to ensure that the version of Scala libraries used by Spark and that custom deployed are same.

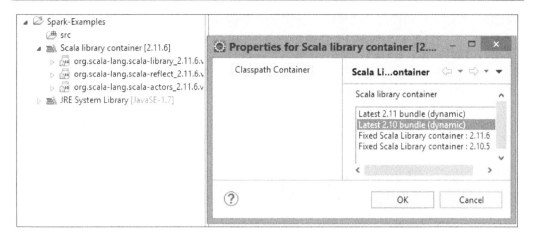

3. Open the properties of your project Spark-Examples and add the dependencies for the all libraries packaged with the Spark distribution, which can be found at $SPARK_HOME/lib.

4. Next, create a Scala package chapter.one and within this package define a new Scala object by the name of ScalaFirstSparkExample.

5. Define a main method in the Scala object and also import SparkConf and SparkContext:

```
package chapter.one

import org.apache.spark.{SparkConf, SparkContext}

object ScalaFirstSparkExample {

  def main(args: Array[String]){
    //Scala Main Method
  }
}
```

6. Now, add the following code to the main method of ScalaFirstSparkExample:

```
    println("Creating Spark Configuration")
    //Create an Object of Spark Configuration
    val conf = new SparkConf()
    //Set the logical and user defined Name of this Application
    conf.setAppName("My First Spark Scala Application")
    //Define the URL of the Spark Master.
    //Useful only if you are executing Scala App directly
    //from the console.
```

Installing and Configuring Spark and Spark Streaming

```scala
//We will comment it for now but will use later
//conf.setMaster("spark://ip-10-237-224-94:7077")

println("Creating Spark Context")
//Create a Spark Context and provide previously created
//Object of SparkConf as an reference.
val ctx = new SparkContext(conf)

println("Loading the Dataset and will further process it")

//Defining and Loading the Text file from the local
//file system or HDFS
//and converting it into RDD.
//SparkContext.textFile(..) - It uses the Hadoop's
//TextInputFormat and file is
//broken by New line Character.
//Refer to http://hadoop.apache.org/docs/r2.6.0/api/org/apache/hadoop/mapred/TextInputFormat.html
//The Second Argument is the Partitions which specify
//the parallelism.
//It should be equal or more then number of Cores in
//the cluster.
val file = System.getenv("SPARK_HOME")+"/README.md";
val logData = ctx.textFile(file, 2)

//Invoking Filter operation on the RDD.
//And counting the number of lines in the Data loaded
//in RDD.
//Simply returning true as "TextInputFormat" have
//already divided the data by "\n"
//So each RDD will have only 1 line.
val numLines = logData.filter(line => true).count()

//Finally Printing the Number of lines.
println("Number of Lines in the Dataset " + numLines)
```

And we are done! Our first Spark program is ready for execution.

Follow the comments provided before each line to understand the code. The same style has been used for all other code examples given in this book.

7. Now from Eclipse itself, export your project as a `.jar` file, name it `Spark-Examples.jar` and save this `.jar` file in the root of `$SPARK_HOME`.

8. Next, open your Linux console, browse to `$SPARK_HOME`, and execute the following command:

   ```
   $SPARK_HOME/bin/spark-submit --class chapter.one.
   ScalaFirstSparkExample --master spark://ip-10-180-61-254:7077
   Spark-Examples.jar
   ```

    We will talk about `spark-submit` at length in the next section but ensure that the value given to parameter `--master` is the same as it is shown on your Spark UI.

9. As soon as you click on *Enter* and execute the preceding command you will see lot of activity (log messages) on the console and finally you will see the output of your job at the end:

[15]

Wow! Isn't that simple! All credit goes to Scala and Spark.

As we move forward and discuss Spark more, you would appreciate the ease of coding and simplicity provided by Scala and Spark for creating, deploying and running jobs in a distributed framework.

Your completed job will also be available for viewing at the Spark UI:

Workers							
Worker Id			Address	State	Cores	Memory	
worker-20150427033250-ip-10-180-61-254.ec2.internal-37691			ip-10-180-61-254.ec2.internal:37691	ALIVE	2 (0 Used)	6.3 GB (0.0 B Used)	

Running Applications							
Application ID	Name	Cores	Memory per Node	Submitted Time	User	State	Duration

Completed Applications							
Application ID	Name	Cores	Memory per Node	Submitted Time	User	State	Duration
app-20150427034150-0000	My First Spark Scala Application	2	512.0 MB	2015/04/27 03:41:50	ec2-user	FINISHED	6 s

The preceding image shows the status of our first Scala job on the UI. Now let's move forward and use Java to develop our Spark job.

Coding Spark jobs in Java

Perform the following steps to write your first Spark example in Java which counts the number of lines given in a text file:

1. Open your `Spark-Example` project created in the previous section and create a new Java file called `JavaFirstSparkExample` in the package `chapter.one`.

2. Define a `main` method in `JavaFirstSparkExample` and also import `SparkConf` and `SparkContext`.

   ```
   package chapter.one

   import org.apache.spark.SparkConf;
   import org.apache.spark.api.java.JavaRDD;
   import org.apache.spark.api.java.JavaSparkContext;
   import org.apache.spark.api.java.function.Function;

   public class JavaFirstSparkExample {

     public static void main(String args[]){
       //Java Main Method
     }
   }
   ```

Now add the following code to the main method of
JavaFirstSparkExample:

```java
System.out.println("Creating Spark Configuration");
// Create an Object of Spark Configuration
SparkConf javaConf = new SparkConf();
// Set the logical and user defined Name of this Application
javaConf.setAppName("My First Spark Java Application");
// Define the URL of the Spark Master
//Useful only if you are executing Scala App directly
//from the console.
//We will comment it for now but will use later
//conf.setMaster("spark://ip-10-237-224-94:7077");

System.out.println("Creating Spark Context");
// Create a Spark Context and provide previously created
//Objectx of SparkConf as an reference.
JavaSparkContext javaCtx = new
JavaSparkContext(javaConf);
System.out.println("Loading the Dataset and will
further process it");

//Defining and Loading the Text file from the local
//filesystem or HDFS
//and converting it into RDD.
//SparkContext.textFile(..) - It uses the Hadoop's
//TextInputFormat and file is
//broken by New line Character.
//Refer to
//http://hadoop.apache.org/docs/r2.6.0/api/org/apache
//hadoop/mapred/TextInputFormat.html
//The Second Argument is the Partitions which specify
//the parallelism.
//It should be equal or more then number of Cores in
//the cluster.
String file = System.getenv("SPARK_HOME")+"/README.md";
JavaRDD<String> logData = javaCtx.textFile(file);

//Invoking Filter operation on the RDD.
//And counting the number of lines in the Data loaded
//in RDD.
//Simply returning true as "TextInputFormat" have
//already divided the data by "\n"
//So each RDD will have only 1 line.
```

```
    long numLines = logData.filter(new Function<String,
    Boolean>() {
      public Boolean call(String s) {
        return true;
      }
    }).count();

    //Finally Printing the Number of lines
    System.out.println("Number of Lines in the
    Dataset "+numLines);

    javaCtx.close();
```

3. Next, compile the preceding `JavaFirstSparkExample` class from Eclipse itself and perform steps 7, 8 and 9 of the previous section in which we executed the Spark Scala example.

And we are done! Analyze the output on the console it should be same as we saw while running the Spark application in Scala.

> We can also execute our Spark jobs locally where we can set the master URL in our Spark jobs to `SparkConf().setMaster("local[2]")` and can execute it as a normal Scala program. Ensure that Spark libraries (`$SPARK_HOME/lib/*.jar`) are in classpath while running the Spark jobs.

In this section we have introduced the basic terminology used in the context of Spark and also written our first Spark program in Java/Scala and further executed the same on the Spark cluster. Let's move forward and see more details about the tools and utilities packaged with the Spark distribution and how they can help us in managing our cluster and jobs.

Tools and utilities for administrators/developers

In this section we will discuss the common tools and utilities available with core Spark packages which can help administrators or developers in managing and monitoring Spark clusters and jobs.

Spark is packaged into 4-5 different folders and each folder has its own significance. Let's move forward and explore a few of these folders and their significance to developers and administrators.

Cluster management

Cluster management is a process of managing and configuring various services, or sets of services provided by a distributed software to form a farm or group of nodes to serve the needs of the user and act as a single unit. It includes various activities like adding and replacing and configuring nodes, scheduling and monitoring jobs or nodes and many more. In this section we will talk about the various utilities available with Spark, which are useful in configuring our Spark cluster:

- `$SPARK_HOME/sbin`: This folder contains all the scripts which help administrators in starting and stopping the Spark cluster. For example: `stop-all.sh` stops all the services with respect to the Spark cluster and `start-all.sh` starts all services (master/slaves) and brings up the complete cluster but, before we use these scripts, we need to create a `slaves` file in the `$SPARK_HOME/conf` folder which will contain the name of all the independent machines on which we wish to execute the Spark workers.

  All the slave nodes should be accessible from master node and `password less ssh` should be configured on all the machines (http://tinyurl.com/l8kp6v3).

   If `password less ssh` doesn't work then you can specify `SPARK_SSH_FOREGROUND` as an environment variable and the scripts will provide you the option to specify the password for each slave in same order as it is mentioned in the `conf/slaves` file.

- `$SPARK_HOME/conf`: This folder contains all the templates for configuring the Spark cluster. The Spark cluster uses the default configurations but developers and administrators can customize them by adding specific configurations and removing `.template` from the filenames. Let's see the usage of different configuration files:

    - `slaves.template`: It is used to define the domain name of the hosts which are entrusted to host Spark workers.
    - `log4.properties.template`: Defines the logging information, which is by default in `INFO` mode. We can customize and provide fine-grained loggers.
    - `spark-defaults.conf.template`: Defines the default Spark configurations used when executing the `$SPARK_HOME/spark-submit` command (see the next section for `spark-submit`).

- `spark-env.sh.template`: Defines the environment variables used by Spark driver/master and worker processes.
- `metrics.properties.template`: This file is used for monitoring purposes where we can configure different metrics provided by the master/worker or driver processes.
- `fairscheduler.xml.template`: Defines the type and mode of scheduler for the Spark jobs.

> Refer to https://spark.apache.org/docs/latest/configuration.html for complete configuration parameters for the Spark master and worker.

Submitting Spark jobs

In this section we will talk about the utilities for submitting our jobs or client programs to our Spark cluster.

The `$SPARK_HOME/bin` folder contains utilities for running or submitting the Spark jobs to the Spark cluster. We have already seen the usage of `spark-class` and `spark-submit`. `spark-class` represents the base driver for running any custom Scala or Java code on the Spark cluster while `spark-submit` provides additional features like launching applications on YARN/Mesos, querying job status, killing jobs, and so on.

Another utility which is worth mentioning is `spark-shell`. This utility creates a SparkContext and provides a console where you can write and directly submit your Spark jobs in Scala. Here is the exact syntax for `spark-shell`:

`$SPARK_HOME/bin/spark-shell -master <url of master>`

`spark-shell` is helpful in debugging Spark jobs where developers want to write and check the output of each line interactively.

Troubleshooting

In this section we will talk about tips and tricks which are helpful when solving the most common errors encountered while working with Spark.

Configuring port numbers

Spark binds various network ports for communication and exposing information to developers and administrators. There may be instances where the default ports used by Spark may not be available or may be blocked by the network firewall which in turn will result in modifying the default Spark ports for master/worker or driver.

Here is the list of all ports utilized by Spark and their associated parameters, which need to be configured for any changes http://spark.apache.org/docs/latest/security.html#configuring-ports-for-network-security.

Classpath issues – class not found exception

Spark runs in a distributed model as does the job. So if your Spark job is dependent upon external libraries, then ensure that you package them into a single JAR file and place it in a common location or the default classpath of all worker nodes or define the path of the JAR file within SparkConf itself. It should look something like this:

```
val sparkConf = new SparkConf().setAppName("myapp").setJars(<path of Jar file>))
```

Other common exceptions

In this section we will talk about few of the common errors/issues/exceptions encountered by developers when they set up Spark or execute Spark jobs.

Setting up Spark clusters and executing Spark jobs is a seamless process but, no matter what we do, there may be errors or exceptions which we see while working with Spark. The following are a few such exceptions and resolutions which should help users in troubleshooting:

- **Too many open files**: Increase the ulimit on your Linux OS by executing sudo ulimit -n 20000.
- **Version of Scala**: Spark 1.3.0 supports Scala 2.10, so if you have multiple versions of Scala deployed on your box, then ensure that all versions are same, that is, Scala 2.10.
- **Out of memory on workers in standalone mode**: Configure SPARK_WORKER_MEMORY in "$SPARK_HOME/conf/spark-env.sh. By default it provides total memory - 1G to workers but, at the same time, you should analyze and ensure that you are not loading or caching too much data on worker nodes.

- **Out of memory in applications executed on worker nodes**: Configure `spark.executor.memory` in your `SparkConf`, like this:

  ```
  val sparkConf = new SparkConf().setAppName("myapp")
  .set("spark.executor.memory", "1g")
  ```

The preceding tips will help you solve basic issues in setting up Spark clusters but, as you move ahead, there could be more complex issues which are beyond the basic setup and for all those issues please post your queries at http://stackoverflow.com/questions/tagged/apache-spark or mail at user@spark.apache.org.

Summary

Throughout this chapter, we have gone through the various concepts and installation procedures of Spark and its various other components. We have also written our first Spark job in Scala and Java and executed the same in distributed mode. At the end we also discussed solutions for fixing common problems encountered during the setup of the Spark cluster.

In the next chapter, we will talk about the architecture of Spark and Spark Streaming and will also write and execute Spark Streaming examples.

2
Architecture and Components of Spark and Spark Streaming

Apache Hadoop brought a revolution in data processing and storage space when it enabled fault-tolerant and distributed processing of large data (TBs/PBs) over commodity machines. Developed on the MapReduce programing model (http://en.wikipedia.org/wiki/MapReduce), Hadoop provided a low cost solution and reliable batch processing (http://en.wikipedia.org/wiki/Batch_processing) of large data.

Hadoop was a perfect fit for most of the varied and complex use cases but there was still a large set of use cases like real-time data processing and computations, iterative data processing (machine learning), and graph processing which were not possible with Hadoop and were still a distant dream for architects and developers, mainly due to two reasons:

- Excessive and intensive use of disks for all intermediate stages
- Only provides map and reduce operations and no other operations like joining/flattening, and grouping of datasets.

And that's where Apache Spark was introduced as a general-purpose data processing engine which can be deployed on variety of infrastructures like YARN, Mesos, and standalone too.

Spark takes MapReduce to the next level with enabled in-memory data storage and near real-time data processing. At the same time, it introduced lot of new operators for performing different kinds of operations such as joins, merging, grouping and many more. Spark introduced a new framework where applications directly run in the systems memory (RAM) and are 100 times faster and 10 times faster even when running on disk as compared to the same applications running on Hadoop cluster.

The architecture of Spark is quite different from other computing frameworks where it not only supports a variety of programming languages such as Java, Scala, Python and R but also provides libraries and extensions built upon Spark Core APIs:

- SQL for structured data processing
- MLlib for iterative data processing—machine learning
- GraphX for graph processing
- Spark Streaming—real-time data processing of streaming data

This chapter will help you understand the overall architecture and various components of Spark and Spark Streaming. It will also help you to develop your first Spark Streaming example in Scala and Java. At the end of this chapter, you will be able to comprehend and appreciate the architecture of Spark and will be ready to take a deep dive and explore some real world use cases of Spark Streaming.

This chapter will cover the following points:

- Batch versus real-time data processing
- Architecture of Spark
- Architecture of Spark Streaming
- Your first Spark Streaming program

Batch versus real-time data processing

In this section we will talk about the complexities involved in working with batch and real-time data processing. We will also discuss factors differentiating data processing of large datasets into batch or real time. This will help us in understanding and appreciating the Spark architecture defined in subsequent sections.

Let's move forward and understand batch and real-time data processing.

Batch processing

Batch processing is a process of defining a series of jobs which are connected with each other or executed after one other in a sequence or in parallel; finally, the output of all the jobs is consolidated to produce the final output. In batch processing, input data is collected in batches over a period of time and the output of one batch can be the input of another. It is also called non-continuous processing of data which is comprised of a collection of large data files (GBs/TBs/PBs) and fast response time is not critical to the business.

Every batch job is associated with the batch window, which is refereed as the time window for processing the jobs and a period of less intensive online activity, as prescribed by production enterprise constraints.

In most cases, batch jobs are scheduled and run at predefined intervals or at a specific time or on a certain day in a month or year.

A few examples of batch jobs are:

- **Log analysis**: In this application, logs are collected over a period of time (per day, week or month) and analysis is performed for deriving various **KPI – key performance indicators** for the underlying system (http://en.wikipedia.org/wiki/Log_analysis)
- **Billing applications**: Calculating the usage of a service provided by the vendor over a period of time and generating billing information, for example, credit companies producing billing information at the end of each month
- **Backups**: A series of jobs running at non-critical business hours and taking the backup of the critical systems
- **Data warehouses**: Consolidating business information in aggregated and static views like weekly/month/yearly reports

And the list goes on…

The preceding examples may foster a perception that batch jobs are not critical in nature, which is an incomplete and false statement.

Batch jobs are critical to business though instant response is not expected from the batch jobs. For example, an offer recommendation job which analyses past data and generates recommendations for a user for upcoming offers can be executed every night and compute the data and store it some persistent storage. But it needs to finish the computation within the stipulated time and, if doesn't, users would not see updated offers/recommendations, which in turn will impact the business.

Let's talk briefly about the complexity involved in batch processing systems:

- **Large data**: Data is really huge and requires a good amount of computational resources to produce results in a fixed time window.
- **Scalability**: This need to have a scale-out architecture so that it can meet the demands of growing data just by adding some more computational resources without re-architecting the complete solution.
- **Distributed processing**: There is a limit to the computational resources (CPUs/RAM) that can be added to a single machine and eventually, as your data grows, your batch jobs may run out of resources. Batch jobs should support scale-out architecture: http://en.wikipedia.org/wiki/Scalability#Horizontal_and_vertical_scaling where they should support distributed processing and computations of data over the cluster of servers.
- **Fault tolerant**: Failures do happen and, especially when your job is running over multiple severs, there could be failures. There could be hardware or network failure or many other external factors which are not under the control of the user/system. For example, a few nodes in the cluster went down while they were processing the batch jobs. So your system should be fault-tolerant in such a manner that it should resume the processing of the failed steps in a job and should not reprocess everything.
- **Enterprise constraints**: Enterprises in today's world impose constraints like SLAs where the batch jobs needs to be completed in a stipulated time so that the system resources can be used for other purposes. There are many more such enterprise constraints like job scheduling, reprocessing, clustering, persistence and so on.

The preceding descriptions should be sufficient to understand that designing or architecting batch applications is not simple. It requires knowledge to develop a distributed, scalable and performance-efficient system for processing batch jobs efficiently and effectively.

We will talk shortly about how Spark handles all these complexities but, before that, let's move forward and discuss real-time data processing and its complexities.

Real-time data processing

Real-time data processing is receiving constantly changing data, such as information relating to air-traffic control, travel booking systems, and so on, and processes it sufficiently rapidly to be able to control the source of the data. The response time for processing data in real time is instant and expected to be in milliseconds (sometimes microseconds too).

A system is said to be a real-time data processing system only when it produces the logically correct results within the given time frame (milliseconds) and if they miss their SLAs or deadlines then there are consequences.

Real-time systems frequently use the term **latency,** which is the time interval between the stimulation and response, or in our world it is the difference between the time the data was received and the response generated. The lesser, the better.

Real-time data processing is often also referred to as near real-time because of the latency introduced or the relaxation in the SLAs in producing the desired results.

Here are a few examples of real-time systems which receive data in near or real-time, process it and send back the results:

- Bank ATMs: Receive an input from the user and instantly reflect the transactions (withdrawal or any other request) to the centralized account.
- Real-time monitoring: Capturing and analyzing data emitted from various data sources like sensors, logs live feeds, and so on in real time.
- Real-time business intelligence: Process of delivering **business intelligence (BI)** or information about business operations as they occur. For more information, refer to `https://en.wikipedia.org/wiki/Real-time_business_intelligence`.
- **Operational intelligence** (OI): It uses real-time data processing and CEP (`http://en.wikipedia.org/wiki/Complex_event_processing`) to gain insight into operations by running query analyses against live feeds and event data. For more information, refer to `http://en.wikipedia.org/wiki/Operational_intelligence`.
- **Point of Sale (POS) systems**: Update inventory, provide inventory history, and sales of a particular item allowing an organization to run payments in real time.
- **Assembly lines**: Process data in real time to reduce time, cost and errors. Errors are instantly captured and appropriate actions are taken without any delays which could otherwise have produced low quality or faulty products.

And the list goes on…

Let's talk briefly about the complexity involved in real-time data processing systems:

- **System responsiveness**: The un-said expectations from real-time data processing systems are that they should process data as it arrives within milliseconds or microseconds, and should not introduce any delays in producing results.

- **Fault-tolerant**: Failures do happen but real-time systems cannot afford to lose a single event.
- **Scalable**: This need to have a scale-out architecture so that it can meet the demands of growing data by adding more computational resources without re-architecting the complete solution.
- **In memory**: Real-time systems cannot afford to read/write from disks, so data processing needs to be handled within the memory itself. So the systems should ensure sufficient memory for storing the input data in the system's memory.

The preceding descriptions should be sufficient to understand that, like batch processing, designing or architecting, real-time data processing system is not so simple.

There were systems like Apache Hadoop (https://hadoop.apache.org/) or Storm (https://storm.apache.org/) used either for batch or real time but not for both. It was acceptable but both systems offered a different programing paradigm which was difficult to maintain. It also added complexities for application developers when they had to follow two different architectural styles of designing systems for batch and real time, which ideally should have been absorbed and handled by the underlying framework itself.

As we always say, necessity is the mother of invention, and that's what happened. Apache Spark was developed as a next generation framework and a one-stop solution for all use cases irrespective of the fact of whether they needed to be processed in batches or real time.

Let's move towards the next section and dive into the architecture of Spark and Spark Streaming and see how they handle the complexities of batch processing and real-time data processing systems.

Architecture of Spark

In this section we will discuss the need for the Spark framework in comparison to Hadoop and then we will also talk about the architecture of Spark which is also referred as its Core Spark Framework.

Spark versus Hadoop

Apache Spark is an open source cluster computing framework which seemed to be similar to Apache Hadoop but actually it is superior to Hadoop. Hadoop performed well for the majority of large scale and distributed data processing over commodity boxes but it failed in two scenarios:

- Iterative and interactive computations and workloads: For example, machine learning algorithms which reuse intermediate or working datasets across multiple parallel operations.
- Real-time data processing: Hadoop was mainly built for batch processing where it lacks in-memory data processing capabilities which are necessary for real-time data processing.

> Refer to https://www.typesafe.com/blog/how-spark-beats-mapreduce-event-streaming-iterative-algorithms-and-elasticity for more information on the pitfalls of Hadoop for iterative and streaming computations.

The preceding scenarios were the motivating factors for a system which could efficiently and effectively support iterative as well as interactive data processing.

The creators of Apache Spark realized that they not only needed to retain the benefits of Apache Hadoop like batch computations, scalability, fault tolerance and distributed data processing but it also needed a new architecture which avoided expensive reads from the disks during various map/reduce stages and supported in-memory processing of distributed data over the cluster of nodes.

Apache Spark does exactly the same thing and introduced a new layer abstraction of distributed datasets which is partitioned over the set of machines (cluster) and can be cached in the memory to reduce the latency, which they named **RDD** (**Resilient Distributed Datasets**).

RDD, by definition, is an immutable (read-only) collection of objects partitioned across a set of machines that can be rebuilt if a partition is lost.

Before moving forward it is important to mention that Spark is capable of performing in-memory operations but it can work on-disk at the same time too.

Spark is designed to be a general execution engine and whenever the data does not fit in memory, Spark interacts with the underlying data storage layer to transfer data to the permanent storage and free up memory.

Let's move forward and discuss the architecture of Spark and its approach of implementing in-memory and disk-based distributed data processing over clusters of nodes.

Layered architecture – Spark

Spark provides a well-defined and layered architecture where all its layers and components are loosely coupled, and integration with external components and extensions is performed using well-defined contracts. Spark architecture also provides the flexibility to define custom libraries and extensions by extending its core APIs.

Here is the high-level architecture of Spark and its various layers which also show the deployment stack of a typical Spark production deployment:

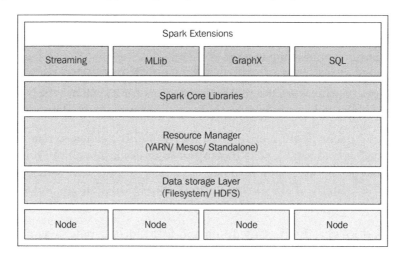

The preceding illustration shows the high-level architecture of Spark which is divided into the following layers:

- **Data storage layer**: This layer is responsible for providing the persistent storage area to Spark applications so that Spark workers can dump data whenever the memory is not sufficient. Spark is extensible and capable of using any kind of filesystem. RDDs, which hold the data, are agnostic to the underlying storage layer and can persist the data in various persistent storage areas like local filesystems, HDFS or any other NoSQL database like HBase, Cassandra, MongoDB, S3, Elasticsearch.
- **Resource manager APIs**: These APIs are used to allocate the available resources (across the cluster) to the client jobs and once the job is finished these resources are reclaimed.

- **Spark Core libraries**: The Spark Core library contains APIs providing the Spark general execution engine which runs on the Spark platform, providing in-memory distributed data processing and a generalized execution model to support a wide variety of applications and languages.
- **Spark extensions/libraries**: These are additional functionalities which are developed by extending the Spark Core APIs to support different use cases. For example, Spark Streaming is one such extension which is developed for performing computations over real-time and streaming data.

Let's re-define and arrange the components which we defined in the *Configuring and running the Spark cluster* section of *Chapter 1*, *Installing and Configuring Spark and Spark Streaming*, and analyze how it fits into the different layers of the Spark architecture.

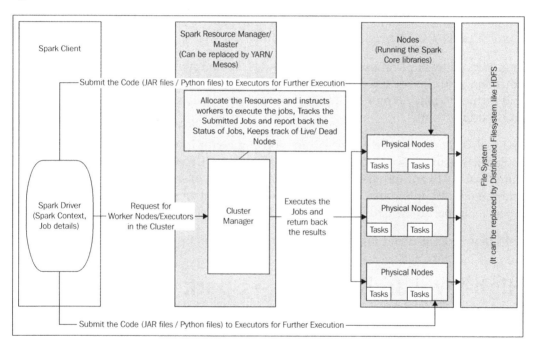

The preceding illustration shows the interaction between the different layers of the Spark architecture. We are using the core Spark libraries, which provide a general execution engine for in-memory distributed data processing.

We will discuss one of the most popular Spark extensions for real-time data processing of streaming data in the next section—Spark Streaming. The rest of the chapters will strenuously examine Spark Streaming using real-world examples and we will also introduce its other extensions.

Let's move forward and talk about the architecture of Spark Streaming and write our first program which consumes the streaming data and does computations in real time.

Architecture of Spark Streaming

In the previous section we discussed Spark, highlighting the position of Spark Streaming in the overall architecture of Spark. In this section we will discuss the various components and architecture of Spark Streaming.

What is Spark Streaming?

Spark Streaming is an interesting and powerful extension of Spark which provides the processing of streaming data or fast moving data (http://en.wikipedia.org/wiki/Stream_(computing)) in near real-time.

There are many applications and use cases like spam filtering, intrusion detection, and clickstream data analysis which generates live data in milliseconds and further to make it more complex it needs to be analyzed at the same time and produce results.

Spark already provided a general execution engine for in-memory data processing, so Spark Streaming was introduced as an extension of Spark which leveraged the same Spark API and provides the processing upon live/streaming data in near real-time. It also integrates with a variety of popular data sources including HDFS, Apache Flume, Apache Kafka, Twitter and TCP Sockets.

Let's move forward and understand the architecture of Spark Streaming.

High-level architecture – Spark Streaming

Spark Streaming implemented the concept of micro-batching where the live/streaming data is divided into a series of deterministic micro-batches and each batch is processed as an individual record and the further output of each batch is sent to the user-defined output streams and can be further persisted into HDFS, NoSQL or can be used to create live dashboards. We will talk more about persistence in *Chapter 5, Persisting Log Analysis Data*.

The size of each batch is governed by the acceptable latency stated by the underlying use case. It can be a few milliseconds or seconds. For example, it collects all Twitter feeds every 100 millisecond and processes them as one single batch. So the system will create a series of multiple batches each containing the feeds received within 100 milliseconds and will submit them to Spark for further processing as a continuous process.

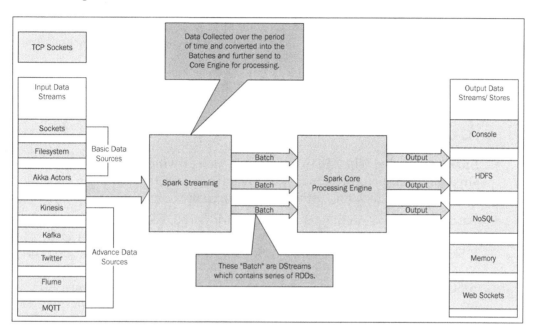

The preceding illustration shows the high-level architecture of Spark Streaming.

Let's discuss in detail the various components defined in the preceding architecture:

- **Input data streams**: These are input data sources which are generating data at a very high velocity (every second, millisecond or even less) and sending it in the form of a continuous stream of data. Spark provides certain connectors to connect to these incoming data streams and capture the data for further processing, which are further categorized into basic and advanced data sources:
    - **Basic data sources**: The data connectors or APIs for these data sources are in-built and packaged with core Spark Streaming bundles. We do not need any external libraries for these connectors.

- **Advanced data sources**: The data connectors or APIs for advanced sources are not part of core Spark Streaming bundles. They require linking or the inclusion of external libraries as dependencies in your build scripts or may be directly downloaded from the community website at http://spark-packages.org/?q=tags%3A%22Streaming%22 and include in the classpath of your application. You can also extend and build your own custom receivers/extensions. Refer to http://spark.apache.org/docs/latest/streaming-custom-receivers.html for building custom receivers.

> Refer to the link for a list of all advanced data at https://spark.apache.org/docs/1.3.0/streaming-programming-guide.html#linking.

- **Spark Streaming**: The API or extension which provides the consumption of streaming data at a predefined interval and converts the streaming data into series batches which are further sent to the Spark Core engine for processing.
- **Batch**: Batches are nothing more than a series of RDDs. Spark Streaming provides another layer of abstraction over a series of RDDs known as **DStreams (Discretized streams)**. DStreams hold the reference of a series of RDDs which are based on the data directly provided by input streams or processed data streams generated by transforming the input streams. All operations on DStreams are applied directly to the underlying RDDs. We will discuss DStreams and its associated operations more in *Chapter 3, Processing Distributed Log Files in Real Time*.
- **Spark Core engine**: The core engine of Spark which receives the input in the form of RDD and further processes and finally sends it to the associated output streams for storage.
- **Output data streams**: The output of each processed batch is directly sent to the output streams for further actions. These output streams can be of varied types, ranging from a raw file system, NoSQL, Queues or web sockets for visualizing the streaming data.

Let's see this working and move on to the next section where we will write our first Spark Streaming program in Scala and Java.

Your first Spark Streaming program

In this section we will code and deploy our first Spark Streaming job in Scala and Java.

Our streaming job will connect to a specific port number and will receive or capture data at regular intervals. It will evaluate and calculate the count of distinct words in the data received. Lastly, it will print this on the console.

The Spark Streaming example shown in subsequent sections is comprised of two distinct parts:

- **Spark Streaming job**: The Spark Streaming job, which contains the actual application logic and will be executed on the Spark cluster
- **Client application**: The client application which opens a specific port and will write some data to that port at regular intervals

Let's move forward and see the code for the Spark Streaming job in Scala and Java and then we will also see the client application. Finally, we will deploy everything on the cluster and will see the results on the console.

Coding Spark Streaming jobs in Scala

Let's extend our Scala project Spark-Examples which we created in *Chapter 1, Installing and Configuring Spark and Spark Streaming*, perform the following steps and code a Spark Streaming example in Scala which counts the number of distinct words received from the client:

1. Create a Scala package chapter.two and, within this package, define a new Scala object called ScalaFirstStreamingExample.
2. Define a main method in the Scala object and also import the packages as shown in the following example:

   ```
   package chapter.two

   import org.apache.spark.SparkConf
   import org.apache.spark.streaming.StreamingContext
   import org.apache.spark.streaming._
   import org.apache.spark.storage.StorageLevel._
   import org.apache.spark.rdd.RDD
   import org.apache.spark.streaming.dstream.DStream
   import org.apache.spark.streaming.dstream.ForEachDStream

   object ScalaFirstStreamingExample {
   ```

Architecture and Components of Spark and Spark Streaming

```
      def main(args: Array[String]){
        //Scala Main Method
      }
    }
```

3. Now add the following code to the main method of
   `ScalaFirstStreamingExample`:

   ```
   println("Creating Spark Configuration")
   //Create an Object of Spark Configuration
   val conf = new SparkConf()
   //Set the logical and user defined Name of this
   Application
   conf.setAppName("My First Spark Streaming Application")

   println("Retreiving Streaming Context from Spark Conf")
   //Retrieving Streaming Context from SparkConf Object.
   //Second parameter is the time interval at which
   //streaming data will be divided into batches
   val streamCtx = new StreamingContext(conf, Seconds(2))

   //Define the type of Stream. Here we are using TCP
   //Socket as text stream,
   //It will keep watching for the incoming data from a
   //specific machine (localhost) and port (9087)
   //Once the data is retrieved it will be saved in the
   //memory and in case memory
   //is not sufficient, then it will store it on the Disk
   //It will further read the Data and convert it into DStream
   val lines = streamCtx.socketTextStream("localhost",
   9087, MEMORY_AND_DISK_SER_2)

   //Apply the Split() function to all elements of DStream
   //which will further generate multiple new records from
   //each record in Source Stream
   //And then use flatmap to consolidate all records and
   //create a new DStream.
   val words = lines.flatMap(x => x.split(" "))

   //Now, we will count these words by applying a using map()
   //map() helps in applying a given function to each
   //element in an RDD.
   val pairs = words.map(word => (word, 1))
   ```

```
//Further we will aggregate the value of each key by
//using/applying the given function.
val wordCounts = pairs.reduceByKey(_ + _)

//Lastly we will print all Values
//wordCounts.print(20)

myPrint(wordCounts,streamCtx)
//Most important statement which will initiate the
//Streaming Context
streamCtx.start();
//Wait till the execution is completed.
streamCtx.awaitTermination();
```

4. Now define one more function `printValues(...)` in your `ScalaFirstStreamingExample`:

```
/**
 * Simple Print function, for printing all elements of RDD
 */
def printValues(stream:DStream[(String,Int)],streamCtx:
StreamingContext){
    stream.foreachRDD(foreachFunc)
    def foreachFunc = (rdd: RDD[(String,Int)]) => {
      val array = rdd.collect()
      println("---------Start Printing Results----------")
      for(res<-array){
        println(res)
      }
      println("---------Finished Printing Results----------
      ")
    }
}

//Most important statement which will initiate the
//Streaming Context
streamCtx.start();
//Wait till the execution is completed.
streamCtx.awaitTermination();
```

And we are done! Our first streaming job in Scala is ready but before we deploy this on the Spark cluster, let's see the same implementation in Java and also code the client application which will send the data to our streaming job.

Coding Spark Streaming jobs in Java

Perform the following steps to code the Spark Streaming example in Java which counts the number of words received on a specific port number:

1. Create a Java file `JavaFirstStreamingExample` in the package `chapter.two` of `Spark-Examples`.

2. Define a main method in the Java class and also import the packages, as shown in the following example:

   ```
   package chapter.two;

   import java.util.Arrays;

   import org.apache.spark.*;
   import org.apache.spark.api.java.function.*;
   import org.apache.spark.storage.StorageLevel;
   import org.apache.spark.streaming.*;
   import org.apache.spark.streaming.api.java.*;

   import scala.Tuple2;

   public class JavaFirstStreamingExample {

     public static void main(String[] s){
       //Java Main Method
   }
   ```

3. Now add the following code to the main method of `JavaFirstStreamingExample`:

   ```
   System.out.println("Creating Spark Configuration");
   //Create an Object of Spark Configuration
   SparkConf conf = new SparkConf();
   //Set the logical and user defined Name of this Application
   conf.setAppName("My First Spark Streaming
   Application");
   //Define the URL of the Spark Master.
   //Useful only if you are executing Scala App directly
   //from the console.
   //We will comment it for now but will use later
   //conf.setMaster("spark://ip-10-237-224-94:7077")

   System.out.println("Retreiving Streaming Context from
   Spark Conf");
   ```

```java
//Retrieving Streaming Context from SparkConf Object.
//Second parameter is the time interval at which
//streaming data will be divided into batches
JavaStreamingContext streamCtx = new
JavaStreamingContext(conf, Durations.seconds(2));

//Define the type of Stream. Here we are using TCP
//Socket as text stream,
//It will keep watching for the incoming data from a
//specific machine (localhost) and port (9087)
//Once the data is retrieved it will be saved in the
//memory and in case memory
//is not sufficient, then it will store it on the Disk.
//It will further read the Data and convert it into DStream
JavaReceiverInputDStream<String> lines =
streamCtx.socketTextStream("localhost",
9087,StorageLevel.MEMORY_AND_DISK_SER_2());

//Apply the x.split() function to all elements of
//JavaReceiverInputDStream
//which will further generate multiple new records from
//each record in Source Stream
//And then use flatmap to consolidate all records and
//create a new JavaDStream.
JavaDStream<String> words = lines.flatMap( new
FlatMapFunction<String, String>() {
    @Override public Iterable<String> call(String x) {
        return Arrays.asList(x.split(" "));
    }
});


//Now, we will count these words by applying a using
//mapToPair()
//mapToPair() helps in applying a given function to
//each element in an RDD
//And further will return the Scala Tuple with "word"
//as key and value as "count".
JavaPairDStream<String, Integer> pairs =
words.mapToPair(
    new PairFunction<String, String, Integer>() {
        @Override
        public Tuple2<String, Integer> call(String s)
        throws Exception {
            return new Tuple2<String, Integer>(s, 1);
        }
```

```
    });


    //Further we will aggregate the value of each key by
    //using/applying the given function.
    JavaPairDStream<String, Integer> wordCounts =
    pairs.reduceByKey(
        new Function2<Integer, Integer, Integer>() {
        @Override public Integer call(Integer i1, Integer
        i2) throws Exception {
            return i1 + i2;
        }
    });


    //Lastly we will print First 10 Words.
    //We can also implement custom print method for
    //printing all values,
    //as we did in Scala example.
    wordCounts.print(10);
    //Most important statement which will initiate the
    //Streaming Context
    streamCtx.start();
    //Wait till the execution is completed.
    streamCtx.awaitTermination();
```

And we are done! Our first streaming job in Java is ready. Let's move on to the next section where we will create our client application which will send the data to our streaming job.

The client application

The client application will allow the user to type messages on the console and capture data. Once data is captured, it will immediately send it to the specific port number (socket) where our streaming job is waiting for the data.

Perform the following steps to create the client application:

1. Extend the Eclipse project Spark-Examples and create a Java class ClientApp.java in the package chapter.two.
2. Define the main method and add imports as shown in the following code:
   ```
   package chapter.two;

   import java.net.*;
   ```

```
import java.io.*;
public class ClientApp {

    public static void main(String[] args) {
       //Main Method
    }
}
```

3. Next add the following piece of code in the main method:

```
try{
    System.out.println("Defining new Socket");
    ServerSocket soc = new ServerSocket(9087);
    System.out.println("Waiting for Incoming Connection");
    Socket clientSocket = soc.accept();

    System.out.println("Connection Received");
    OutputStream outputStream =
    clientSocket.getOutputStream();
    //Keep Reading the data in a Infinite loop and send it
    //over to the Socket.
    while(true){
        PrintWriter out = new PrintWriter(outputStream,
        true);
        BufferedReader read = new BufferedReader(new
        InputStreamReader(System.in));
        System.out.println("Waiting for user to input some
        data");
        String data = read.readLine();
        System.out.println("Data received and now writing
        it to Socket");
        out.println(data);

    }

}catch(Exception e ){
    e.printStackTrace();
}
```

And we are done with our client application!

Now let's move on to the final step where we will package and deploy our Spark Streaming job on the Spark cluster.

Packaging and deploying a Spark Streaming job

The packaging and deployment of Spark Streaming jobs on the Spark cluster is very similar to the process followed for deploying standard Spark batch jobs but with one additional step, in which we need to ensure that our data stream providing the data is up and running before our job is deployed.

Perform the following steps for packaging and deploying a streaming job:

1. We will export your project as a `.jar` file, name it `Spark-Examples.jar` and save this `.jar` file in the root of `$SPARK_HOME`. Execute the following command on your Linux console, from the location where you compiled your `Spark-Examples` project:

   ```
   jar -cf $SPARK_HOME/Spark-Examples.jar *
   ```

2. Next open your Linux console, browse to `$SPARK_HOME`, and execute the following command for initializing your client application:

   ```
   java -classpath Spark-Examples.jar chapter.two.ClientApp
   ```

   Once your client app is up and running it will wait till someone (in our case it will be the streaming job) is there to receive the data. Let's refer to this console as `ClientConsole`:

   ```
   sumit@localhost $ java -classpath Spark-Examples.jar chapter.two.ClientApp
   Defining new Socket
   Waiting for Incoming Connection
   ```

3. Next open another Linux console, browse to `$SPARK_HOME`, and execute the following command for deploying your Spark Streaming job:

   ```
   $SPARK_HOME/bin/spark-submit --class chapter.two.ScalaFirstStreamingExample --master <SPARK-MASTER-URL> Spark-Examples.jar
   ```

4. Let's refer to this console as `SparkJobConsole`.

5. As soon as you click on *Enter* and execute the preceding command you should see lot of activity (log messages) on the `SparkJobConsole` and your streaming job is ready to receive the data from your `ClientApp` every two seconds. Whatever you type in your `ClientConsole`, you will see the count of those words appearing in your `SparkJobConsole`.

6. For example, type some text in your `ClientConsole` window, as shown in the following illustration:

Chapter 2

```
sumit@localhost $ java -classpath Spark-Examples.jar chapter.two.ClientApp
Defining new Socket
waiting for Incoming Connection
Connection Received
waiting for user to input some data
This is my First Spark Streaming Example... Soon we will see more Such Examples ...
Data received and now writing it to Socket
waiting for user to input some data
```

7. You will see the count of the distinct words in the text sent by our `ClientApp` on the `SparkJobConsole`:

```
15/05/12 01:27:44 INFO BlockManagerInfo: Added broadcast_15_piece0 in memory on ip-10-37-204-59.ec2.internal:38429 (size: 1573.0 B, free: 264.9 MB)
15/05/12 01:27:44 INFO MapOutputTrackerMasterActor: Asked to send map output locations for shuffle 12 to sparkExecutor@ip-10-37-204-59.ec2.internal:54236
15/05/12 01:27:44 INFO MapOutputTrackerMaster: Size of output statuses for shuffle 12 is 159 bytes
15/05/12 01:27:44 INFO MapOutputTrackerMasterActor: Asked to send map output locations for shuffle 12 to sparkExecutor@ip-10-37-204-59.ec2.internal:44906
15/05/12 01:27:44 INFO TaskSetManager: Finished task 1.0 in stage 26.0 (TID 95) in 54 ms on ip-10-37-204-59.ec2.internal (1/2)
15/05/12 01:27:44 INFO TaskSetManager: Finished task 0.0 in stage 26.0 (TID 94) in 63 ms on ip-10-37-204-59.ec2.internal (2/2)
15/05/12 01:27:44 INFO TaskSchedulerImpl: Removed TaskSet 26.0, whose tasks have all completed, from pool
15/05/12 01:27:44 INFO DAGScheduler: Stage 26 (foreachRDD at Scala_FirstStreamingExample.scala:61) finished in 0.064 s
15/05/12 01:27:44 INFO DAGScheduler: Job 13 finished: foreachRDD at Scala_FirstStreamingExample.scala:61, took 0.135590 s
---------Start Printing Results---------
(,...,2)
(Example,1)
(is,1)
(will,1)
(we,1)
(This,1)
(first,1)
(my,1)
(Spark,1)
(Soon,1)
(Examples,1)
(such,1)
(more,1)
(see,1)
(Streaming,1)
---------Finished Printing Results---------
15/05/12 01:27:44 INFO JobScheduler: Finished job streaming job 1431394064000 ms.0 from job set of time 1431394064000 ms
15/05/12 01:27:44 INFO JobScheduler: Total delay: 0.147 s for time 1431394064000 ms (execution: 0.137 s)
15/05/12 01:27:44 INFO ShuffledRDD: Removing RDD 42 from persistence list
```

The preceding example shows a stateless computation or a count of words received from the user every two seconds.

8. You will be able to see the status of your streaming job on Spark Master UI under the **Running Applications** tab:

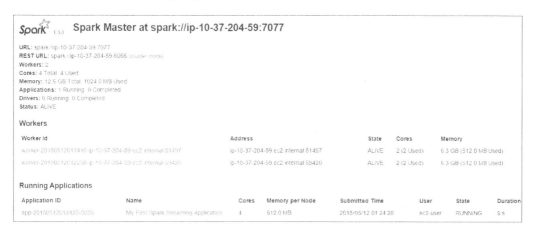

Wow! Isn't that simple and easy! That is the real power of Spark—one framework for all your batch and streaming jobs.

[43]

In this section, we have developed and deployed our first Spark Streaming job, which counts the number of distinct words from the text sent by client applications over a socket in real time.

Let's move on to the next chapter where we will see the complex real world examples of Spark Streaming.

Summary

In this chapter, we have discussed the challenges and programming paradigms for batch and real-time data processing. We also discussed the need for frameworks like Spark and its differences with preexisting frameworks like Hadoop. At the end, we developed and deployed our first Spark Streaming program.

In the next chapter, we will talk about Spark Client APIs and some of its integral components. We will also introduce a real-world use case for distributed data processing in real time.

3
Processing Distributed Log Files in Real Time

In today's world, large amounts of valuable data are stored in repositories distributed across large-scale networks which can be accessed over the Web. The key challenge is to provide a distributed, efficient, scalable, extensible fault-tolerant system to manipulate this data easily, safely, and with high performance.

There is no doubt that systems like Hadoop brought revolution and provided a framework for processing data at Internet scale—large and distributed data in TBs/PBs. But it was primarily meant for data-intensive operations, where it processed data over clusters of nodes/machines and devoted most of its processing time to I/O and manipulation of data.

Apache Spark took distributed computing to the next level when it introduced a new architectural paradigm for data-intensive and computer-intensive operations and also provided extensions like Spark Streaming for processing streaming data in real-time or near real-time.

Spark Streaming is an interesting extension to the core Spark APIs. It provides the distributed processing (loading, manipulating and persisting) of a continuous stream of data while retaining all the other core features of Spark.

Additionally, it provides high-level operators in Java, Scala, and Python for running ad hoc operations on the streaming data enabling developers to build powerful interactive applications.

This chapter will discuss the Spark Streaming APIs for performing various data loading operations. We will deep dive into the core components of Spark and Spark Streaming and, at the end, we will also discuss real-life use cases for distributed log file analysis in real time.

This chapter will cover the following points:

- Spark packaging structure and client APIs—Java and Scala
- Resilient distributed datasets and discretized streams
- Data loading from distributed and varied sources

Spark packaging structure and client APIs

In this section, we will discuss Spark's packaging structure and various client APIs provided by Spark for data loading, manipulation and finally caching in memory or persisting to external storage. We will also talk about the Spark packaging structure which will help us to understand the purpose and functionalities provided by the various packages and classes.

Spark is written in Scala (http://www.scala-lang.org/) but for interoperability it also provides the equivalent APIs in Java and Python.

For brevity we will only talk about the Scala and Java APIs and, for Python APIs, users can refer to https://spark.apache.org/docs/1.3.0/api/python/index.html.

As discussed in *Chapter 2, Architecture and Components of Spark and Spark Streaming*, Spark provides layered architecture and, at a high level, Spark is divided into two different modules:

- **Spark Core**: It contains the core packages and classes for various basic operations like task scheduling/distribution, tracking, I/O operations, memory management, and many more. Spark Core also defines the RDD which are immutable objects and represent the main programming extensions.
- **Spark libraries/extensions**: They contain the various modules built over the Spark Core APIs for providing specialized features. As of Spark 1.3.0, the following extensions are available and packaged with the Spark standard distribution:
    - **Spark Streaming**: Spark Streaming is used for the consuming, transforming, processing and storing of high-throughput live data streams in fault-tolerant manner. It is also known as a real-time or near-real-time extension of Spark for processing streaming data.

- **Spark SQL**: This is a module for structured data processing. Spark SQL provides another layer of abstraction known as DataFrames which is merely a distributed collection of data organized into columns referred by the user defined names. DataFrame are equivalent to the tables in relational databases or frames in Python (http://pandas.pydata.org/pandas-docs/dev/generated/pandas.DataFrame.html) or R (http://www.r-tutor.com/r-introduction/data-frame).

  We will read more about Spark SQL in *Chapter 6, Integration with Advanced Spark Libraries*.

- **Spark MLlib**: This provides machine learning (http://en.wikipedia.org/wiki/Machine_learning algorithms) for statistical analysis including algorithms classification, regression, clustering, collaborative filtering and dimensionality reduction. Spark MLlib provides high-level APIs for defining the sequence of stages or the series of steps executed in order, which is known as machine learning pipelines.

- **Spark GraphX**: Another extension of Spark which exposes APIs for defining and manipulating data using graphs (http://en.wikipedia.org/wiki/Graph_(mathematics)). This extends the RDDs and provides an abstraction for working with directed multigraphs with properties attached to each vertex and edge. It also provides various graph algorithms like PageRank, connected components and triangle counting. GraphX not only provides the functionality of graphs but also retains the core feature of Spark where all graph computations are executed in parallel which is referred as **graph-parallel computations**.

  We will talk more about Spark GraphX in *Chapter 6, Integration with Advanced Spark Libraries*.

> Spark has also defined one more extension, Spark Bagel, but it will soon be replaced by GraphX.

Let's dive into each of the Spark modules and discuss the important components, packages and their associated functionalities.

Spark Core

Spark Core is the heart of Spark and provides two basic components, SparkContext and Spark Config. Both of these components are used by each and every standard or customized Spark job or Spark library and extension. **Context** and **Config** is not a new term/concept and more or less it has now become a standard Architectural Pattern. By definition – a Context is an entry point of the application which provides the access to various resources/features exposed by the framework while the Config contains the application configurations, which helps in defining the environment of the application.

Let's move on to the nitty-gritty of the Scala and Java APIs exposed by Spark Core.

SparkContext and Spark Config – Scala APIs

org.apache.spark.SparkContext is the first statement in any Spark job which defines the SparkContext and then further defines the business logic. SparkContext is the entry point for accessing any of the Spark features which we may want to use or leverage, for example, connecting to the Spark cluster, submitting jobs, and so on. Even the references to all Spark extensions are provided by SparkContext. There can be only one SparkContext per JVM which needs to be stopped if we want to create a new one. SparkContext is immutable, which means it cannot be changed or modified once it is started.

In order to create the SparkContext you need to provide the reference of org.apache.spark.SparkConf which holds the values of all spark.* environment variables defined as Java system properties. There are default values associated with each and every configuration variable, which can be overwritten by explicitly defining it in your Spark job. For example, let's assume that our job is dependent on some custom JAR files and we want to distribute these custom JAR files to the worker nodes before our job is executed, which if not done may produce failures. It can be achieved by the following code snippet:

```
val conf = new SparkConf();
val jars = Seq("$SPARK_HOME/lib/a.jar","$SPARK_HOME/lib/b.jar")
conf.setJars(jars)
val context = new SparkContext(conf)
.........
.......
```

> Refer to https://spark.apache.org/docs/1.3.0/api/scala/index.html#org.apache.spark.SparkContext and https://spark.apache.org/docs/1.3.0/api/scala/index.html#org.apache.spark.SparkConf for complete list of methods exposed by SparkContext and SparkConf.

SparkContext and Spark Config – Java APIs

Though Scala code is eventually converted into class files and executed on JVM itself, there are some APIs which are only defined in Scala and do not have an equivalent in Java. So for all those Scala objects, Spark provides compatible and equivalent Java APIs for interoperability in the org.apache.spark.api.java.* package. For example, org.apache.spark.api.java.JavaSparkContext is defined as a Java-friendly version of SparkContext which provides the same functionality but utilizes the Java primitives instead of Scala.

We need to remember that there are only certain Scala functions which do not have an equivalent in Java like higher order functions (a function within a function), so we will use Java implementations only for those functions and the rest will still use the Scala APIs. For example, org.apache.spark.SparkConf does not have an equivalent Java API, as it already defines Java-friendly or compatible operations.

Further in the book, we will use the Scala APIs for all our examples.

RDD – Scala APIs

org.apache.spark.rdd.RDD.scala is another important component of Spark which represents the distributed collection of datasets. It also exposes various operations which can be executed in parallel over the cluster. SparkContext exposes various methods to load the data from HDFS or the local filesystem or Scala collections and finally create an RDD on which various operations such as map, filter, join, and persist can be invoked.

RDD also defines some useful child classes within the org.apache.spark.rdd.* package like PairRDDFunctions for working with key/value pairs, SequenceFileRDDFunctions for working with Hadoop sequence files and DoubleRDDFunctions for working with RDDs of doubles.

Refer to https://spark.apache.org/docs/1.3.0/api/scala/index.html#org.apache.spark.rdd.RDD for a complete list of methods exposed by RDD.scala and refer to https://spark.apache.org/docs/1.3.0/api/scala/index.html#org.apache.spark.rdd.package for various types of RDDs provided by Spark.

We discussed a few examples in *Chapter 1, Installing and Configuring Spark and Spark Streaming* and *Chapter 2, Architecture and Components of Spark and Spark Streaming* which load data from the local filesystem and create an RDD. Let us see one more example for creating RDDs from the data in relational databases where we will use `org.apache.spark.rdd.JDBCRDD.scala`:

```
//Define the Configuration
val conf = new SparkConf();
//Define Context
val ctx = new SparkContext(conf)

//Define JDBC RDD
val rdd = new JdbcRDD(
ctx,
() => { DriverManager.getConnection("jdbc:derby:temp/Jdbc-RDDExample")
},
    "SELECT EMP_ID,Name FROM EMP WHERE Age > = ? AND ID <= ?",20,
30, 3,
    (r: ResultSet) => { r.getInt(1); r.getString(2) } ).cache()

//Print only first Column in the ResultSet
System.out.println(rdd.first)
```

The preceding example assumes that you have the Apache Derby database with an EMP table with columns — ID, name, age and the ID of all employees within the age group of 20 to 30 and finally print the first record.

> For other available operations, please refer to `JdbcRDD.scala` (https://spark.apache.org/docs/1.3.0/api/scala/index.html#org.apache.spark.rdd.JdbcRDD).

RDD – Java APIs

`org.apache.spark.api.java.JavaRDD` and `org.apache.spark.api.java.JavaRDDLike` are top two classes defined in Scala for interoperability with the Java APIs.

It also defines `JavaDoubleRDD`, `JavaHadoopRDD`, `JavaNewHadoopRDD` and `JavaPairRDD` in `org.apache.spark.api.java.package`.

> Refer to https://spark.apache.org/docs/1.3.0/api/scala/index.html#org.apache.spark.api.java.package for further details on the Java-specific RDD APIs.

Other Spark Core packages

Spark provides many other classes and distributes them as part of Spark Core packages. Let's discuss the important ones which we will be using frequently while developing Spark jobs:

- `org.apache.spark`: It is the core package of the Spark API which contains functionality for creating/distributing/submitting Spark jobs on the cluster.

> For more information, refer to https://spark.apache.org/docs/1.3.0/api/scala/index.html#org.apache.spark.package.

- `org.apache.spark.annotation`: Contains the annotations which are used within the Spark API. This is the internal Spark package and you may not use the annotations defined in this package while developing your custom Spark jobs. The three main annotations defined within this package are:

    - `DeveloperAPI`: All those methods which are marked with `DeveloperAPI` are for advance usage where users are free to extend and modify the default functionality. These methods may be changed or removed in the next minor or major releases.

    - `Experimental`: All those methods or classes which are officially not adopted by Spark but are introduced temporarily in a specific release are marked as `Experimental`. These methods may be changed or removed in the next minor or major releases.

    - `AlphaComponent`: All those methods or classes which are still in the testing phase and are not recommended for production use are marked as `AlphaComponent`. These methods may be changed or removed in the next minor or major releases.

- `org.apache.spark.broadcast`: Provides the API for sharing the read-only variables across the Spark jobs. Once the variables are defined and broadcast, they cannot be changed. This is one of the important packages which are frequently used by developers in their custom Spark jobs. Broadcasting the variables and data across the cluster is a complex task and we need to ensure that an efficient mechanism is used so that it improves the overall performance of the Spark job and does not become an overhead. Spark provides two different types of implementations of broadcasts— `HttpBroadcast` and `TorrentBroadcast`. `HttpBroadcast` leverages the HTTP server as a broadcast mechanism. In this mechanism the broadcast data is fetched from the driver, through a HTTP Server running at the driver itself and further stored in the executor Block Manager for faster accesses. `TorrentBroadcast`, which is also the default implementation of the broadcast, maintains its own Block Manager. The first request to access the data makes the call to its own Block Manager and, if not found, the data is fetched in chunks from the executor or driver. Torrent Broadcast works on the principle of BitTorrent and ensures that the driver is not the bottleneck in fetching the shared variables and data.

> Spark also provides accumulators which work like broadcast but provide updatable variables shared across the Spark jobs but with some limitations. Please refer to https://spark.apache.org/docs/1.3.0/api/scala/index.html#org.apache.spark.Accumulator.

- `org.apache.spark.io`: Provides implementation of various compression libraries which can be used at block storage level. This whole package is marked as `DeveloperAPI`, so developers can extend and provide their own implementations. By default, it provides three implementations—LZ4, LZF and Snappy. Again developers may not use it directly but if required, it can provide custom implementations, for performance-tuning the compression/decompression mechanism.

> For more details, refer to https://spark.apache.org/docs/1.3.0/api/scala/index.html#org.apache.spark.io.package.

- `org.apache.spark.scheduler`: This provides various scheduler libraries which help in job scheduling, tracking and monitoring. It defines the **Directed Acyclic Graph (DAG)** scheduler http://en.wikipedia.org/wiki/Directed_acyclic_graph. Spark DAG scheduler defines the stage-oriented scheduling where it keeps track of the completion of each RDD and the output of each stage and then computes DAG which is further submitted to the underlying `org.apache.spark.scheduler.TaskScheduler` that runs them on the cluster.

> Refer to https://spark.apache.org/docs/1.3.0/api/scala/index.html#org.apache.spark.scheduler.package.

- `org.apache.spark.serializer`: Defines the APIs for serialization and deserialization used in data shuffling and RDD.

> Refer to https://spark.apache.org/docs/1.3.0/api/scala/index.html#org.apache.spark.serializer.package.

- `org.apache.spark.storage`: Provides APIs for structuring, managing and finally persisting the data stored with RDD within blocks. It also keeps tracks of data and ensures it is either stored in memory or, if the memory is full, it is flushed to disk.

> Refer to https://spark.apache.org/docs/1.3.0/api/scala/index.html#org.apache.spark.storage.package.

- `org.apache.spark.ui`: Contains the classes containing the data which needs to be displayed on Spark UI.
- `org.apache.spark.util`: Utility classes for performing common functions across the Spark APIs. For example, it defines `MutablePair` which can be used as an alternative to Scala's `Tuple2` with the difference that `MutablePair` is updatable while Scala's `Tuple2` is not. It helps in optimizing memory and minimizing object allocations.

Spark libraries and extensions

Spark extensions are the auxiliary features developed by extending Core Spark APIs. Spark extensions are segregated from the Core Spark API and packaged into their own packages.

Let's see the packaging structure of each of the Spark extensions and discuss a few of the important classes provided by each of the Spark extensions.

Spark Streaming

All Spark Streaming classes are packaged in the `org.apache.spark.streaming.*` package. Spark Streaming defines two critical classes, `StreamingContext.scala` and `DStream.scala`. Let's examine the functions and roles performed by these classes:

- `org.apache.spark.streaming.StreamingContext`: It is similar to `SparkContext` but provides an entry point to Spark Streaming functionality. It defines methods to create the objects of `DStream.scala` from input sources. It also provides functions to start and stop the Spark Streaming jobs.

- `org.apache.spark.streaming.dstream.DStream.scala`: DStream, or discretized streams, provides the basic abstraction of Spark Streaming. It provides the sequence of RDDs created from the live data or transforming the existing DStreams. This class defines the global operations which can be performed on all DStreams and a few specific operations which can be applied on specific types of DStreams.

Spark Streaming also defines various sub packages for providing implementation for various types of input receivers:

- `org.apache.spark.streaming.flume.*`: Provides classes for consuming input data from Fume (https://flume.apache.org/).
- `org.apache.spark.streaming.kafka.*`: Provides classes for consuming input data from Kafka (http://kafka.apache.org/).
- `org.apache.spark.streaming.mqtt.*`: Provides classes for consuming input data from MQTT (http://mqtt.org).
- `org.apache.spark.streaming.kinesis.*`: Provides classes for consuming input data from Amazon Kinesis (http://aws.amazon.com/kinesis/).
- `org.apache.spark.streaming.twitter.*`: Provides classes for consuming input data from Twitter feeds.

- `org.apache.spark.streaming.zeromq.*`: Provides classes for consuming input data from Zero MQ (http://zeromq.org/).

>
> For more information on Spark Streaming APIs, refer to https://spark.apache.org/docs/1.3.0/api/scala/index.html#org.apache.spark.streaming.package.
> Spark Streaming also define compatible Java classes in the `org.apache.spark.streaming.api.java.*` package for interoperability with Java APIs.

Spark MLlib

Spark provides a scalable machine learning library consisting of common machine learning algorithms and utilities. It includes various standard algorithms for classification (http://en.wikipedia.org/wiki/Statistical_classification), regression (http://en.wikipedia.org/wiki/Regression_analysis), clustering (http://en.wikipedia.org/wiki/Cluster_analysis), and collaborative filtering (http://en.wikipedia.org/wiki/Recommender_system#Collaborative_filtering).

All Spark machine learning libraries are packaged in the `org.apache.spark.mllib.*` package. Let's see some of the main packages provided by the `org.apache.spark.mllib.*` package:

- `org.apache.spark.mllib.classificaiton.*`: Provides implementation of various machine learning algorithms for classification like Naive Bayes (http://en.wikipedia.org/wiki/Naive_Bayes_classifier), support vector machines (http://en.wikipedia.org/wiki/Support_vector_machine) and many more. Refer to https://spark.apache.org/docs/latest/mllib-classification-regression.html for more information on supported classification models.

- `org.apache.spark.mllib.recommendation.*`: Provides classes for collaborative filtering which are generally used to build recommended systems. For more details, refer to https://spark.apache.org/docs/latest/mllib-collaborative-filtering.html.

- `org.apache.spark.mllib.clustering.*`: Provides implementation of various clustering algorithms. For more information, please refer to https://spark.apache.org/docs/latest/mllib-clustering.html.

> Spark 1.2 also introduced a new package called org.apache.spark.ml, which is an alpha component and aims to provide a uniform set of high-level APIs for creating and tuning practical machine learning pipelines. For more information, refer to https://spark.apache.org/docs/latest/ml-guide.html.

Spark SQL

Spark SQL provides the processing of structured data and facilitates the execution of relational queries which are expressed in structured query language. (http://en.wikipedia.org/wiki/SQL).

All Spark SQL classes are packaged in org.apcahe.spark.sql.* and its subpackages. Let us discuss a few of the important Spark SQL packages:

- org.apache.spark.sql.*: Topmost package which contains all classes and subpackages for the implementation of Spark SQL.

> For more information, refer to https://spark.apache.org/docs/1.3.0/api/scala/index.html#org.apache.spark.sql.package.

- org.apache.spark.sql.hive.*: Provides the implementation for executing Apache Hive queries over Spark.

> For more information, refer to https://spark.apache.org/docs/1.3.0/api/scala/index.html#org.apache.spark.sql.hive.package.

- org.apache.spark.sql.sources.*: Provides classes and APIs for adding different kinds of data sources to Spark SQL. Spark SQL provides a unified interface for loading and persisting data from various data sources like CSV, JSON, JDBC, Parquet, Hive, AWS S3, Amazon Redshift, HBase, MongoDB, Elasticsearch, Solr, and many more. It also provides the flexibility for extending and adding new data sources. A few of the data source APIs for loading the data from Hive, Parquet and JDBC (MySQL and PostgreSQL) are distributed with the standard Spark distribution while for others we need to download from the vendor or Spark community website and manually configure.

> For more information, refer to https://spark.apache.org/docs/1.3.0/api/scala/index.html#org.apache.spark.sql.sources.package.
>
> Databricks provide the list of third party packages and extensions for various other data sources (http://spark-packages.org/) with an understanding that support will be directly provided by the vendor and not by Spark or Databricks.

Spark GraphX

Spark GraphX is another extension which provides loading, structuring and distributed processing of data in the form of graphs: http://en.wikipedia.org/wiki/Graph_(abstract_data_type).

GraphX supports the **property graph data model** which is a combination of vertices and edges. The graph provides basic operations to access and manipulate the data associated with vertices and edges as well as the underlying structure. Like Spark RDDs, the graph is a functional data structure in which mutating operations return new graphs.

Let's discuss a few of the important Spark GraphX packages:

- org.apache.spark.graphx.*: Topmost package which contains all classes and subpackages for the implementation of GraphX. For more information, refer to https://spark.apache.org/docs/1.3.0/api/scala/index.html#org.apache.spark.graphx.package.

- org.apache.spark.graphx.impl.*: It contains the implementations for a few of the abstract classes defined in org.apache.spark.graphx package like org.apache.spark.graphx.VertexRDD, org.apache.spark.graphx.EdgeRDD or org.apache.spark.graphx.Graph. For more information, refer to https://spark.apache.org/docs/1.3.0/api/scala/index.html#org.apache.spark.graphx.impl.package.

- org.apache.spark.graphx.lib.*: It contains the various utility classes, analytical functions and algorithms for processing graphs like PageRank (http://en.wikipedia.org/wiki/PageRank), SVD (http://en.wikipedia.org/wiki/Singular_value_decomposition) and shortest path (http://en.wikipedia.org/wiki/Shortest_path_problem). For more information, refer to https://spark.apache.org/docs/1.3.0/api/scala/index.html#org.apache.spark.graphx.lib.package.

> Refer to https://spark.apache.org/docs/latest/graphx-programming-guide.html for more information on GraphX.

In this section we discussed the important Spark APIs/classes and packages provided by Spark and its various libraries/extensions. Let's move on and discuss in detail the architecture of RDD and DStreams which are a core abstraction layer for providing the distributed processing of the datasets.

Resilient distributed datasets and discretized streams

In this section we will discuss the architecture, motivation and other important concepts related to resilient distributed datasets. We will also talk about the implementation methodology adopted by Spark libraries/extensions like Spark Streaming for extending and exposing resilient distributed datasets.

Resilient distributed datasets

Resilient distributed datasets (RDD) is an independent concept which was developed in the University of California, Berkeley and was first implemented in systems like Spark to show its real usage and power. RDD is a core component of Spark. It provides in-memory representation of immutable datasets for parallel and distributed processing. RDD are more of abstraction (an agnostic of the underlying data store) providing the core functionality of in-memory data representation for storing and retrieving data objects which can be further extended to capture various data structures like graph or relational structures or streaming data. Let's move forward and talk about the nitty-gritty of RDD.

Motivation behind RDD

Frameworks like Hadoop and MapReduce are widely adopted for parallel and distributed data processing. There is no doubt that these frameworks introduce a new paradigm for distributed data processing and that too in fault-tolerant manner (without losing a single byte). But it does have some limitations. It is not suited for the problem statements where we need iterative data processing as in recursive functions or machine learning algorithms where data needs to be in-memory for the computations.

For all those scenarios, a new paradigm—RDD was introduced which contains all the features of Hadoop like systems like distributed processing, fault-tolerant, and so on but essentially keeps data in the memory providing distributed in-memory data processing over a cluster of nodes.

Let us move forward and discuss the important features of RDDs.

Fault tolerance

One of the main challenges of RDD was to provide an efficient fault-tolerance mechanism for the datasets which are already loaded and processed in the memory. Though this not a new concept and is already being implemented in various distributed processing systems like Hadoop, key-value stores and so on but the architecture used by these systems is not good enough for in-memory distributed data processing. Frameworks like Hadoop, provided fault tolerance by maintaining multiple copies of the same dataset over the various nodes in the cluster or maintain the log of updates happening over the original datasets and applying the same over the machines/nodes. This process is good for disk-based systems but the same mechanism is inefficient for data-intensive workloads or memory-based systems because first they require copying large amounts of data over the cluster network, whose bandwidth is far lower than that of RAM, and second they incur substantial storage overheads.

RDD introduced a new concept for fault tolerance and provided a coarse-grained interface based on transformations. Now, instead of replicating data or keeping logs of updates, RDDs keep track of the transformations (like map, reduce, join, and so on) applied to the particular dataset, which is also called a **data lineage**.

This allows an efficient mechanism for fault tolerance where, in event of loss of any partition, RDD has enough information to derive the same partition by applying the same set of transformations on the original dataset. Moreover this computation is parallel and involves processing on multiple nodes, so the recomputation is very fast and efficient too, in contrast to costly replication used by other distributed data processing frameworks.

Transformations and actions

RDDs are read-only collections of records which are partitioned over the cluster of nodes. They are created by applying deterministic operations over the distributed datasets residing in any persistence storage system like disks, Queues, NoSQL, RDBMS, and so on. These deterministic operations are known as **transformations**. Transformations by design are lazy and are not applied instantly unless the results are requested by the driver program. The operations which apply the transformation and return the value to the drivers are known as **actions**.

For example, invoking a map(...) function is an transformation which does not yield results but invoking reduce(...) is an action which yields results and returns the output to the driver.

A few of the frequently used transformation functions provided by RDDs are:

- map(func): Applies the provided function to each element of the source dataset and generates a new distributed RDD.
- filter(func): Applies the provided function to all the elements of the source datasets and generates a new RDD containing only those elements for which func returned true.
- union(otherDataset): Joins the datasets and returns a new RDD containing all the elements of an RDD on which union was invoked and the elements (RDD) were passed as an argument.
- sortByKey([asc/desc], [numOfTasks]): When called on a dataset of key/value, it returns a dataset of key/value sorted by keys in ascending or descending order, as specified in the first argument. By default, it is ascending.

All of the preceding transformation functions return a new/fresh RDD. Remember, RDDs are immutable and, once created cannot be changed.

A few of the frequently used action functions provided by RDDs are:

- reduce(func): Aggregates all the elements of the datasets by applying func. The provided func should be commutative and associative so that it can be computed correctly in parallel.
- collect(): Returns all the elements of the datasets as an array to the driver program.
- first(): Returns the first element of the RDD or dataset.
- take(n): Returns the first *n* elements of the RDD or dataset.

Refer to https://spark.apache.org/docs/latest/api/scala/index.html#org.apache.spark.rdd.RDD for a complete list of operations provided by RDD.

RDD storage

RDDs by design are distributed and partitioned over clusters of nodes which are kept in the system's memory but, at the same time, it provides operations which can be used to store RDD on disks or to an external system. The native integration for storing RDD in a filesystem and HDFS is provided by Spark Core packages while it is also extended by the community and various other vendors to provide storage of RDD in external storage systems like MongoDB, DataStax, Elasticsearch, and so on.

The following are a few of the functions which can be used for storing RDDs:

- `saveAsTextFile(path)`: Writes the elements of RDD to a text file in a local filesystem or HDFS or any other mapped or mounted network drive.
- `saveAsSequenceFile (path)`: Writes the elements of RDD as a Hadoop Sequence file in a local filesystem or HDFS or any other mapped or mounted network drive. This is available on RDDs of key-value pairs that implement Hadoop's writable interface.
- `saveAsObjectFile(path)`: Write the elements of the dataset to a given path using Java serialization mechanism, which can then be loaded using `SparkContext.objectFile(path)`.

RDD persistence

Persistence in RDD is also called caching of RDD. It can simply be done by invoking `<RDD>.persist(StorageLevel)` or `<RDD>.cache()`. By default, RDD is persisted in memory (default for `cache()`) but it also provides the persistence on disks or any other external systems which are defined by the provided function `persist` and its parameter of class `StorageLevel` (https://spark.apache.org/docs/latest/api/scala/index.html#org.apache.spark.storage.StorageLevel$).

`StorageLevel` is annotated as `DeveloperApi()` which can be extended to provide the custom implementation of persistence.

Caching or persistence is a key tool for iterative algorithms and fast interactive use. Whenever persist is invoked on RDD, each node stores its associated partitions and computes in memory and further reuses them in other actions on the computed datasets. This in turn enables the future actions to be much faster.

Shuffling in RDD

Shuffling redistributes data across clusters so that it is grouped differently across the partitions. It is a costly operation as it involves copying data across executors and nodes and then creating new partitions and it then distributes them across the cluster.

There are certain transformation operations defined in `org.apache.spark.rdd.RDD` and `org.apache.spark.rdd.PairRDDFunctions` which trigger shuffling and repartition. These operations include:

- `RDD.repartition(...)`: Repartitions the existing dataset across the cluster of nodes
- `RDD.coalesce(...)`: Repartitions the existing dataset into a smaller number of given partitions
- All operations which ends by `ByKey` (except count operations) like `PairRDDFunctions.reducebyKey()` or `groupByKey`
- All join operations like `PairRDDFunctions.join(...)` or `PairRDDFunctions.cogroup(...)` operations

Shuffling is a costly and time-consuming operation as it requires the recomputation of datasets which may span partitions, invoking operations like `groupByKey(....)` for grouping, or for performing aggregation like sum or average. It is a real challenge because not all values for a single key necessarily reside on the same partition/node or machine, but they need to be collocated to compute the results. During computations, a single task will operate on a single partition but, to organize all the data for a single `groupByKey` task to execute, it will read from all partitions and find all the values for all keys, and then bring all the values across the partitions to compute the final result for each key.

The shuffle is an expensive operation since it involves disk I/O, data serialization, and network I/O but there are certain configurations which can help in tuning and performance optimizations.

Refer to `https://spark.apache.org/docs/latest/configuration.html#shuffle-behavior` for a complete list of parameters which can be used to optimize the shuffle operations.

Refer to https://www.cs.berkeley.edu/~matei/papers/2012/nsdi_spark.pdf for more information on RDDs.

In this section we have talked about the core and intrinsic features of RDD. Now as we have enough information on RDDs, let's move on and discuss the way it has been extended by Spark extensions like Spark Streaming.

Discretized streams

Discretized streams are also called DStreams and are a new stream processing model in which computations are structured as a series of stateless, deterministic batch computations at small time intervals.

This new stream processing model not only enables powerful recovery mechanisms (similar to those in batch systems) but it also out-performs replication and upstream backup.

DStreams leveraged the concepts of resilient distributed datasets and created a series of RDDs (of the same type) as one single DStream which is processed and computed at a user-defined time interval. DStreams can be created either from input data streams which are connected to data sources like filesystems, sockets, and so on. DStreams can also be created by applying high-level operations on other DStreams. There can be multiple DStreams per Spark Streaming context and each DStream contains a series of RDDs. Each RDD is the snapshot of the data received at a particular point in time from the receiver. The duration or interval after which data needs to be converted or computed into RDDs is defined as the second parameter of `StreamingContext`:

```
val streamCtx = new StreamingContext(conf, Seconds(2))
```

For example, the data from Twitter feeds is collected every 100 milliseconds but it is processed every second:

```
val streamCtx = new StreamingContext(conf, Seconds(1))
```

So the DStream of this feed would look similar to the following illustration:

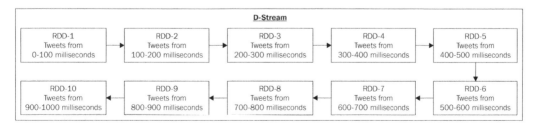

DStreams are primarily an execution strategy, breaking down computation into steps. They implement and use most of the standard operations which are used in other streaming systems like "sliding window" or "incremental processing". The real benefit of DStreams was a Unified Architecture and design for processing Streams and, at the same time, it also exposed Stream operations over other interfaces like SQL.

DStreams support two types of operations:

- **Transformations**: Supports all transformation operations as supported by RDD like `map()`, `flatmap()`, and so on and creates a new DStream from the processed data. Transformations are applied separately on the RDD at each time interval.
- **Output operations**: These operations help to save the final output to external systems or the same external cache. Again it is similar to the output operations supported by RDD.

Apart from all the standard operations, DStreams support two additional operations:

- **Windowing**: Windowing is a special type of operation which is provided only by DStreams and groups all the records from a sliding window of past time intervals into one RDD. We will talk more about windowing functions in *Chapter 4, Applying Transformations to Streaming Data*.
- **Incremental aggregation / stateful processing**: This provides the functionality for a common use case where we may need to compute an aggregate like a count or max over a sliding window. DStreams provide several variants of an incremental operation. All methods in `DStream.scala` prefixed with `Window` provide incremental aggregations like `countByWindow`, `reduceByWindow`, and so on.

Refer to `https://spark.apache.org/docs/latest/api/scala/index.html#org.apache.spark.streaming.dstream.DStream` for more information on the operations supported by the DStream API.

In this section we have talked about the core and intrinsic features of discretized streams for stream processing in Spark. Now let us move forward and discuss the use case where we will use all these concepts and load the streaming data from multiple and varied data sources.

Data loading from distributed and varied sources

In large enterprises, log file analysis is one of the popular use cases. Architects/business analysts and all other stake holders always want to analyze the logs of various activities like events, security, access, and so on and uncover the hidden patterns. For example, the web logs from a popular user interfacing application (a website or portal) can easily provide you with the following information:

- Most popular pages: Frequently visited pages
- Type of browsers or user agent used by consumers to visit the website
- Origin of the user: Users' referrer
- Final status of the user request (HTTP status codes): Successful (200), broken links (404), redirection (301), and many more

Consider another example where enterprises are always concerned about unauthorized access to their network and shared resources. Network administrators used to spend a considerable amount of time in analyzing the access logs or event logs where they were constantly looking for any unauthorized access or suspicious activities with respect to their critical, important and confidential network resources.

There are many more use cases where log analysis can provide insights into existing applications or network resources which not only help in improving the overall user experience (in the case of web applications) but also protect from outside threats like hacking (`http://en.wikipedia.org/wiki/Hacker_(computer_security)`) or spoofing (`http://en.wikipedia.org/wiki/Website_spoofing`).

It becomes more complex when all this information needs to be consumed and analyzed in real time. For example analyzing the network access logs at end of the day does not make sense because the hackers might have completed their job by that time.

Let's explore one such use case for web log analysis where an online system (a website or any portal) is deployed on a couple of servers (clustered deployment) serving the user request from multiple nodes or machines and log files are also generated on each of these nodes, which would look like similar to the following illustration:

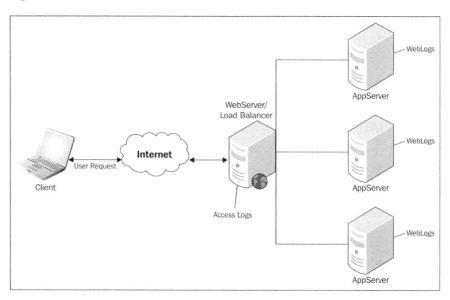

The preceding illustration shows a typical deployment of an online and user interfacing application.

The first challenge with this kind of distributed system is that the weblogs are generated on multiple servers, though logically they are from the same application. In order to solve this challenge, we will use Spark Streaming and its integration with Flume to read the streaming log files from multiple sources and load it in our Spark Streaming application.

> For more information on Flume, refer to https://flume.apache.org/.

Let's discuss the architecture and core components of Flume and then we will set up Flume and write our custom Spark Streaming application and integrate it with Flume.

Flume architecture

Flume is a reliable and highly available service for efficiently and effectively consuming streaming data in near real-time (with negligible latency of milliseconds) from different data sources. It is highly customizable and extendable, robust and provides fault tolerance with tunable reliability mechanisms. Flume is slowly and gradually becoming a de facto standard for performing **ETL** (**Extract, Transform, Load**) for events or data received in real or near real-time. Let's discuss the architecture of Flume and the components it uses for performing ETL.

Flume works in a distributed mode in which it defines following four components:

- **Source**: Source is a component which reads or extracts (**Extract** of ETL Process) the data from a configurable data source. Flume defines a few standard components which can be configured to read the data from predefined data sources like databases, filesystems, sockets, and so on. Every chunk of data read by the Flume source is known as an **event**, which is further delivered to a channel.

> Refer to `https://flume.apache.org/FlumeUserGuide.html#flume-sources` for standard Flume sources and a developer guide and `https://flume.apache.org/FlumeDeveloperGuide.html#source` for developing custom sources.

- **Channel**: Channels are the staging area where events are stored, so that they can be consumed further by Flume sinks. It decouples the sources from the sink so that the sink is not dependent upon the source. There could be scenarios where the source could extract the events or data faster than the events and data consumption capacity of the sink. Channels help the source and the sink to work independently at their own capacity. Channels can be reliable or non-reliable. For example, Flume defines standard channels like memory, JDBC, file, and so on, where the memory channel is unreliable while the file and JDBC are reliable channels.

> Refer to `https://flume.apache.org/FlumeUserGuide.html#flume-channels`, for more information on the standard channels provided by Flume.

- **Interceptor**: Interceptor is another component which modifies, enhances or transforms (**Transform** of ETL process) data before it is consumed by the sink. An interceptor can modify or even drop events based on any criteria chosen by the developer or the interceptor. Flume supports the chaining of interceptors. This is made possible by specifying the list of interceptor builder class names in the configuration.

> Refer to `https://flume.apache.org/FlumeUserGuide.html#flume-interceptors` for standard Flume interceptors.

- **Sink**: Sink is another component which retrieves or loads (**Load** of ETL process) data from channels and then can optionally process or transform and finally deliver it to the intended recipients or consumer or another system like Kafka or Avro or store it to a persistent storage area like files or HDFS, and so on. Similar to the source, Flume also provides standard sinks like HDFS, file, and so on.

Sink completes our pure ETL pipeline where we have the Flume source for extracting, the Flume interceptor for transformation and the Flume sink for loading events into various systems or sending data to a bus or a processing system like Spark or Storm.

> Refer to `https://flume.apache.org/FlumeUserGuide.html#flume-sinks` for standard Flume sinks and a developer guide and `https://flume.apache.org/FlumeDeveloperGuide.html#sink` for developing a custom sink.

All the four Flume components (source, channel, interceptor and sink) work in conjunction and the high-level architecture is same as shown in the following illustration:

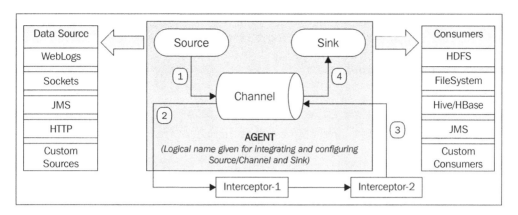

In the next section we will configure Flume for consuming data from web logs.

Installing and configuring Flume

Perform the following steps to configure Flume:

1. Download Flume from `http://www.apache.org/dyn/closer.cgi/flume/1.6.0/apache-flume-1.6.0-bin.tar.gz` and extract it using the following command on the same machine or server where your web log files are generated:

   `tar - zxvf apache-flume-1.6.0-bin.tar.gz`

2. Execute the following command to define the `FLUME_HOME` environment variable:

   `export FLUME_HOME=<path of extracted Flume binaries>`

3. Next, browse the `FLUME_HOME/conf` directory and create a new config file named `spark-flume.conf`.

4. Edit the `spark-flume.conf` file and add following content:

```
#Defining Agent-1 and the logical names of the Source/ Channel and
Sink
a1.sources = src-1
a1.channels = c1
a1.sinks = spark

#Defining Agent-2 and the logical names of the Source/ Channel and
Sink
a2.sources = src-2
a2.channels = c2
a2.sinks = spark1

#Configuring Source for Agent-1
#Here we are defining a source which will execute a custom Linux
Command "tail" to get the Data from configured web log file
a1.sources.src-1.type = exec
#Name of the Log File with the full path
a1.sources.src-1.command = tail -f /home/servers/node-1/
appserver-1/logs/debug.log
#Define the Channel which will be used by Source to deliver the
messages.
a1.sources.src-1.channels = c1

#Defining and providing Configuration of Channel for Agent-1
#Memory channel is not a reliable channel.
a1.channels.c1.type = memory
a1.channels.c1.capacity = 2000
a1.channels.c1.transactionCapacity = 100

#Configuring Sink for Agent-1
a1.sinks.spark.type = spark
#This is the Custom Sink which will be used to integrate with our
Spark Application
a1.sinks.spark.type = org.apache.spark.streaming.flume.sink.
SparkSink
#Name of the host where this Sink is running
a1.sinks.spark.hostname = localhost
#Custom port where our Spark-Application will connect and consume
the event
a1.sinks.spark.port = 4949
```

Chapter 3

```
#Define the Channel which will be used by Sink to receive the
messages.
a1.sinks.spark.channel = c1

#Configuring Source for Agent-2
#Here we are defining a source which will execute a custom Linux
Command "tail" to get the Data from configured web log file
a2.sources.src-2.type = exec
#Name of the Log File with the full path
a2.sources.src-2.command = tail -f /home/servers/node-1/
appserver-2/logs/debug.log
#Define the Channel which will be used by Source to deliver the
messages.
a2.sources.src-2.channels = c2

#Defining and providing Configuration of Channel for Agent-2
a2.channels.c2.type = memory
a2.channels.c2.capacity = 2000
a2.channels.c2.transactionCapacity = 100

#Configuring Sink for Agent-2
a2.sinks.spark1.type = spark
#This is the Custom Sink which will be used to integrate with our
Spark Application
a2.sinks.spark1.type = org.apache.spark.streaming.flume.sink.
SparkSink
#Name of the host where this Sink is running
a2.sinks.spark1.hostname = localhost
#Custom port where our Spark-Application will connect and consume
the event
a2.sinks.spark1.port = 4950
#Define the Channel which will be used by Sink to receive the
messages.
a2.sinks.spark1.channel = c2
```

 Follow the comments provided in the configuration to understand the purpose of each property. The same style is used later in this chapter and book.

In the preceding Flume configuration we defined two agents (a1 and a2) which monitor the log files from two different app servers hosted on the same machine. Each agent is configured with its own source, channel and sink. We have defined the standard source and channel (provided by Flume) but we have configured a custom sink which will help us in integration with our Spark application.

> Exec is not a reliable source and is not recommended when we need strong event delivery semantics or guaranteed delivery. It is better to develop a custom source using Flume SDK. It is recommended to have a cleanup script with an exec command which runs at regular intervals to check process tables for the tail -f command whose parent PID is 1 and kill them manually as they are dead processes.

5. Next, let's download the following additional libraries for our custom sink and add them to $FLUME_HOME/lib/:
   - http://central.maven.org/maven2/org/apache/spark/spark-streaming-flume_2.10/1.3.0/spark-streaming-flume_2.10-1.3.0.jar
   - http://central.maven.org/maven2/org/apache/spark/spark-streaming-flume-sink_2.10/1.3.0/spark-streaming-flume-sink_2.10-1.3.0.jar

6. Our custom sink also has a dependency upon the Spark Core libraries, so copy $SPARK_HOME/lib/spark-assembly-1.3.0-hadoop2.4.0.jar to $FLUME_HOME/lib/.

   Our Flume configuration is completed and finally we are ready to execute our Flume agents.

7. Browse $FLUME_HOME and execute the following commands for running both the agents (a1 and a2):

   ```
   ./bin/flume-ng agent --conf conf --conf-file conf/spark-flume.conf --name a1 &

   ./bin/flume-ng agent --conf conf --conf-file conf/spark-flume.conf --name a2 &
   ```

8. If everything works as planned than your Flume is up and running and agents are consuming events from the app server log files and delivering to the memory channel.

9. If you see any exceptions, check your configuration, especially the path of log file given in `a1.sources.src-1.command` and `a2.sources.src-2.command` and ensure that it does exists and is accessible to your Flume agents.

10. Next let's write our custom Spark application for consuming the messages from both Flume agents and then print them on the console.

> In the absence of real log files you can write a custom Scala/Java class with which can write logs or text in the configured log files or you can download publicly available log files from http://www.monitorware.com/en/logsamples/download/apache-samples.rar.

Configuring Spark to consume Flume events

Perform the following steps to configure your Spark application to receive the events generated by the Flume agents:

1. Download the following JAR files and place them in the `$SPARK_HOME/lib/` directory:
    - http://central.maven.org/maven2/org/apache/spark/spark-streaming-flume_2.10/1.3.0/spark-streaming-flume_2.10-1.3.0.jar
    - http://central.maven.org/maven2/org/apache/spark/spark-streaming-flume-sink_2.10/1.3.0/spark-streaming-flume-sink_2.10-1.3.0.jar

2. The preceding JAR files are the same as those we downloaded and copied in our `$FLUME_HOME/lib/` directory in the previous *Installing and configuring Flume* section.

3. Rename `$SPARK_HOME/conf/spark-defaults.conf.template` as `$SPARK_HOME/conf/spark-defaults.conf`.

4. Edit your `$SPARK_HOME/conf/spark-defaults.conf` file and add the following entries at the end of the file:

    ```
    spark.driver.extraClassPath=$SPARK_HOME/lib/spark-streaming-flume_2.10-1.3.0.jar:$FLUME_HOME/lib/flume-ng-sdk-1.5.2.jar:$SPARK_HOME/lib/spark-streaming-flume-sink_2.10-1.3.0.jar
    spark.executor.extraClassPath=$SPARK_HOME/lib/spark-streaming-flume_2.10-1.3.0.jar:$FLUME_HOME/lib/flume-ng-sdk-1.5.2.jar:$SPARK_HOME/lib/spark-streaming-flume-sink_2.10-1.3.0.jar
    ```

5. The preceding entries ensure that all Spark dependencies for consuming events from Flume agents are in the classpath of our Spark driver and at the same time they are also available to Spark executors which will be executing our Spark application.

6. Next, we will extend our Scala project Spark-Examples which we created in *Chapter 1, Installing and Configuring Spark and Spark Streaming* and add a new package chapter.three and within this package define a new Scala object by the name of ScalaLoadDistributedEvents.scala.

7. Define a main method in the Scala object and import the packages, as shown in the following example:

```
package chapter.three
import org.apache.spark.SparkConf
import org.apache.spark.streaming._
import org.apache.spark.streaming.flume._
import org.apache.spark.storage.StorageLevel
import org.apache.spark.rdd._
import org.apache.spark.streaming.dstream._
import java.net.InetSocketAddress
import java.io.ObjectOutputStream
import java.io.ObjectOutput
import java.io.ByteArrayOutputStream

object ScalaLoadDistributedEvents {

  def main(args:Array[String]){

    println("Creating Spark Configuration")
    //Create an Object of Spark Configuration
    val conf = new SparkConf()
    //Set the logical and user defined Name of this
    //Application
    conf.setAppName("Streaming Data Loading Application")

    println("Retreiving Streaming Context from Spark Conf")
    //Retrieving Streaming Context from SparkConf Object.
    //Second parameter is the time interval at which
    //streaming data will be divided into batches
    val streamCtx = new StreamingContext(conf, Seconds(2))

    //Create an Array of InetSocketaddress containing the
    //Host and the Port of the machines
```

```
//where Flume Sink is delivering the Events
//Basically it is the value of following properties
//defined in Flume Config: -
//1. a1.sinks.spark.hostname
//2. a1.sinks.spark.port
//3. a2.sinks.spark1.hostname
//4. a2.sinks.spark1.port
var addresses = new Array[InetSocketAddress](2);

addresses(0) = new InetSocketAddress("localhost",4949)
addresses(1) = new InetSocketAddress("localhost",4950)

//Create a Flume Polling Stream which will poll the
//Sink the get the events
//from sinks every 2 seconds.
//Last 2 parameters of this method are important as the
//1.maxBatchSize = It is the maximum number of events
//to be pulled from the Spark sink
//in a single RPC call.
//2.parallelism - The Number of concurrent requests
//this stream should send to the sink.
//for more information refer to
//https://spark.apache.org/docs/1.1.0/api/java/
org/apache/spark/streaming/flume/FlumeUtils.html
val flumeStream = FlumeUtils.createPollingStream
(streamCtx,addresses,StorageLevel.
MEMORY_AND_DISK_SER_2,1000,1)

//Define Output Stream Connected to Console for
//printing the results
val outputStream = new ObjectOutputStream(Console.out)
//Invoking custom Print Method for writing Events to
//Console
printValues(flumeStream,streamCtx, outputStream)

//Most important statement which will initiate the
//Streaming Context
streamCtx.start();
//Wait till the execution is completed.
streamCtx.awaitTermination();
}
```

8. Now define one more function `printValues(...)` in `ScalaLoadDistributedEvents`:

```
/**
  * Simple Print function, for printing all elements of RDD
  */
def printValues(stream:DStream
[SparkFlumeEvent],streamCtx: StreamingContext,
outputStream: ObjectOutput){
  stream.foreachRDD(foreachFunc)
  //SparkFlumeEvent is the wrapper classes containing all
  //the events captured by the Stream
  def foreachFunc = (rdd: RDD[SparkFlumeEvent]) => {
    val array = rdd.collect()
    println("---------Start Printing Results----------")
    println("Total size of Events= " +array.size)
    for(flumeEvent<-array){
      //This is to get the AvroFlumeEvent from
      //SparkFlumeEvent
      //for printing the Original Data
      val payLoad = flumeEvent.event.getBody()
      //Printing the actual events captured by the Stream
      println(new String(payLoad.array()))
    }
    println("---------Finished Printing Results--------")
} }
```

And we are done! Our Spark Streaming job is ready to be deployed and executed. Let's move on to the next section where we will perform the final step and deploy our Spark Streaming application.

Packaging and deploying a Spark Streaming job

Perform the following steps to deploy and run your Spark Streaming job:

1. Export your project as a JAR file and name it `Spark-Examples.jar` and save this JAR file in the root of `$SPARK_HOME`.

Chapter 3

2. Next, open your Linux console and browse to $SPARK_HOME and execute the following command to deploy your Spark Streaming job:

   `$SPARK_HOME/bin/spark-submit --class chapter.three.ScalaLoadDistributedEvents --master <SPARK-MASTER-URL> Spark-Examples.jar`

3. As soon as you click on *Enter* and execute the preceding command you should see that all your web logs are being consumed and printed on the console.

Overall architecture of distributed log file processing

To summarize the overall use case, here is what we have developed and deployed:

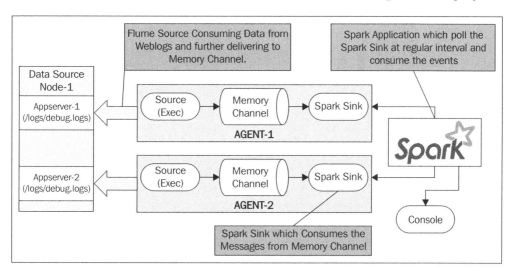

The preceding illustration defines the various components and the overall architecture of our distributed log processing use case. We will continue using the same architecture in subsequent chapters and will exploit and leverage various other features of Spark Streaming.

 Refer to `https://spark.apache.org/docs/1.2.0/streaming-flume-integration.html` for more information on Spark and Flume integration.

Summary

In this chapter, we have discussed in detail the packages and core components and classes of Spark and its extensions. We have also discussed resilient distributed datasets and discretized streams and finally integrated and configured Spark Streaming with Flume and executed our distributed log file processing use case.

In the next chapter, we will discuss and apply various transformation functions over our streaming data and discuss the performance-tuning aspects of our Spark Streaming application/cluster.

4
Applying Transformations to Streaming Data

Data is of no use if it cannot be transformed and some meaningful analysis is derived from the overall process!

Data analysis is a multi-step process which includes inspecting, transforming and finally modeling, with the important goal of discovering useful information which is further considered and applied in arriving at critical business decisions.

In simple terms, transformation is a process where a series of rules or functions is applied to extracted data, so that it can be loaded to the end target for further analysis.

An important activity in transformation is data cleansing, which aims to process and prepare only proper and relevant data, which can be analyzed and interpreted by the system.

Transformation is very subjective and, depending on the end goal, can perform a variety of functions. Aggregations (`sum`, `avg`, `min`, and `max`), sorting, joining data from multiple sources, disaggregations, deriving new values, and so on are some examples of the common functions involved in the transformation process.

In this chapter we will talk about the different transformation functions in Spark Streaming and the process of applying these transformation functions on the distributed streaming data loaded from data sources.

This chapter will cover the following points:

- Understanding and applying transformation functions
- Performance tuning

Understanding and applying transformation functions

Spark Streaming provides various high-level transformation functions which can be used to perform operations such as simple aggregations, counts but also including complex windowing functions over streaming data.

Let's reuse the code snippets written in *Chapter 3, Processing Distributed Log Files in Real Time*, with a different set of log data which will help us to better understand the way these functions are applied to the streaming data.

Simulating log streaming

Perform the following steps for generating Apache access logs and simulate your environment to provide the logs in real time:

1. Download sample Apache access logs from the following location http://www.monitorware.com/en/logsamples/download/apache-samples.rar.

2. Extract the downloaded apache_samples.rar and its subarchives into a folder. Let's refer to this folder as $LOG_FOLDER.

3. Next, we will write the program which will read the Apache access_log at certain intervals and will extract each row and put the log data into the log files monitored by our Flume agents.

4. Create a package called chapter.four and a Java class SampleLogGenerator.java and add the following code to it:

    ```
    package chapter.four;

    import java.io.BufferedReader;
    import java.io.File;
    import java.io.FileOutputStream;
    import java.io.FileReader;

    public class SampleLogGenerator {

      public static void main(String[] args) {
        try {

          if (args.length != 2) {
            System.out.println("Usage - java SampleLogGenerator
            <Location of Log File to be read> <location of the
            log file in which logs needs to be updated>");
    ```

```
            System.exit(0);
        }
        String location = args[0];

        File f = new File(location);
        FileOutputStream writer = new FileOutputStream(f);

        File read = new File(args[1]);
        BufferedReader reader = new BufferedReader(new
        FileReader(read));

        for (;;) {

          writer.write((reader.readLine()+"\n").getBytes());
          writer.flush();
          Thread.sleep(500);
        }

      } catch (Exception e) {
        e.printStackTrace();
      }

  }

}
```

5. Next, edit $FLUME_HOME/conf/spark-flume.conf and change the location of property of a1.sources.src-1.command to /home/servers/node-1/appserver-1/logs/access.log.

6. Next, we will compile the preceding Java program and execute it by providing the following two runtime parameters:

   ```
   java chapter.four.SampleLogGenerator  $LOG_FOLDER/access_log/access_log /home/servers/node-1/appserver-1/logs/access.log
   ```

7. Now you will see that the logs are being generated in /home/servers/node-1/appserver-1/logs/access.log which simulates the real environment where logs are being generated for every request.

8. Next, kill all your Flume agents if you have any running (by executing kill -9 <PID>) and browse $FLUME_HOME and execute the following command to bring up only one Flume agent for consuming the data and delivering it to the Spark sink:

   ```
   ./bin/flume-ng agent --conf conf --conf-file conf/spark-flume.conf --name a1
   ```

We are done with our setup. Now the logs are being generated in real time and we need to consume them and perform further analysis.

Let us move to the next section where we will discuss the transformation operations provided by Spark Streaming.

Functional operations

Discretized streams or DStream are nothing but a series of RDD which are represented by `org.apache.spark.streaming.dstream.DStream.scala` and `org.apache.spark.streaming.dstream.PairDStreamFunctions.scala`. It defines various higher-order functions like map and reduce that accepts and apply a given function to each element of the RDD and produces a new RDD. As per Wikipedia, higher-order functions are those functions that either accept a function as input or output a function. It is a common concept, defined in functional programming languages like Clojure, Lisp, Erlang, Haskell and now in Java-8 too.

The following are the different higher-order functions and operations provided by DStream:

- `flatMap(flatMapFunc): DStream[U]` — Similar to `map(...)` but, before returning, it flattens the results and then returns the final result set.
- `forEachRDD(forEachFunc)`: Applies the given function to all RDDs in a given stream. It is a special type of function and it is worth noting that the given function is applied to all RDDs on the driver node itself but there could be actions defined in the RDD and all those actions are performed over the cluster. `forEach` is categorized as an output operator and, by default, output operations are executed one-at-a-time, sequentially in the order they are defined in the application.
- `filter(filterFunc): DStream[T]` — Applies the provided function to all the elements of RDD and generates the RDD only for those elements which return TRUE.
- `map(mapFunc): DStream[U]` — Applies a given function `mapFunc` to all elements of RDD and generates a new RDD.
- `mapPartitions(mapPartFunc, preservePartitioning)`: Return a new DStream in which each RDD is generated by applying `mapPartitions()` to each RDD in the invoking DStream. Applying `mapPartitions()` to an RDD applies the given function to each partition of the RDD.

Let's continue our log streaming example and see a few of the preceding functions in action.

Perform the following steps and create a custom Spark program for consumption and analysis of logs generated by our Flume agents:

1. We will extend our Spark-Examples project and create a new Scala class ScalaLogAnalyzer.scala in the package, chapter.four.

2. Edit ScalaLogAnalyzer.scala and add the following piece of code:

```
package chapter.four

import java.util.regex.Pattern
import java.util.regex.Matcher

class ScalaLogAnalyzer extends Serializable{

  /**
   * Transform the Apache log files and convert them into a
   Map of Key/Value pair
   */
  def tansfromLogData(logLine: String):Map[String,String]
  ={
    //Pattern which will extract the relevant data from
    Apache Access Log Files
     val LOG_ENTRY_PATTERN = """^(\S+) (\S+) (\S+)
     \[([\w:/]+\s[+\-]\d{4})\] "(\S+) (\S+) (\S+)" (\d{3})
     (\S+)""";
     val PATTERN = Pattern.compile(LOG_ENTRY_PATTERN);
     val matcher = PATTERN.matcher(logLine);

    //Matching the pattern for the each line of the Apache
    access Log file
    if (!matcher.find()) {
      System.out.println("Cannot parse logline" + logLine);
    }
    //Finally create a Key/Value pair of extracted data and
    return to calling program
    createDataMap(matcher);

  }
```

[83]

```
/**
 * Create a Map of the data which is extracted by
   applying Regular expression.
 */
def createDataMap(m:Matcher):Map[String,String] = {
  return Map[String, String](
    ("IP" -> m.group(1)),
    ("client" -> m.group(2)),
    ("user" -> m.group(3)),
    ("date" -> m.group(4)),
    ("method" -> m.group(5)),
    ("request" -> m.group(6)),
    ("protocol" -> m.group(7)),
    ("respCode" -> m.group(8)),
    ("size" -> m.group(9))
  )}
```

The preceding class applies the regular expression (pattern) and transforms a given line of Apache access logs into a map of key/value pairs. This is required so that we can capture each distinct section of the Apache access log statements and then perform further analysis.

3. Next, we will define another Scala object in the package chapter.four and name it ScalaTransformLogEvents.scala, which will consume the log events and apply the different functions to perform analysis.

4. Edit ScalaTransformLogEvents.scala and add the following piece of code which consumes the logs and converts them into key/value pairs:

```
package chapter.four

import org.apache.spark.SparkConf
import org.apache.spark.streaming._
import org.apache.spark.streaming.flume._
import org.apache.spark.storage.StorageLevel
import org.apache.spark.rdd._
import org.apache.spark.streaming.dstream._
import java.net.InetSocketAddress
import java.io.ObjectOutputStream
import java.io.ObjectOutput
import java.io.ByteArrayOutputStream

object ScalaTransformLogEvents {
```

```scala
  def main(args:Array[String]){

    /** Start Common piece of code for all kinds of
    Transform operations*/
    println("Creating Spark Configuration")
    val conf = new SparkConf()
    conf.setAppName("Apache Log Transformer")
    println("Retreiving Streaming Context from Spark Conf")
    val streamCtx = new StreamingContext(conf, Seconds(10))

    var addresses = new Array[InetSocketAddress](1);
    addresses(0) = new InetSocketAddress("localhost",4949)

    val flumeStream =
    FlumeUtils.createPollingStream(streamCtx,addresses,
    StorageLevel.MEMORY_AND_DISK_SER_2,1000,1)

    //Utility class for Transforming Log Data
    val transformLog = new ScalaLogAnalyzer()
    //Invoking Flatmap operation to flatening the results
    and convert them into Key/Value pairs
    val newDstream = flumeStream.flatMap { x =>
    transformLog.tansfromLogData(new
    String(x.event.getBody().array())) }

    /** End Common piece of code for all kinds of Transform
    operations*/

    /**Start - Transformation Functions */
    executeTransformations(newDstream,streamCtx)
    /**End - Transformation Functions */

    streamCtx.start()
    streamCtx.awaitTermination()
  }

   /**
    * Define and execute all Transformations to the log
    data
    */
  def
  executeTransformations(dStream:DStream[(String,String)],
  streamCtx: StreamingContext){ }
}
```

Most of the preceding piece of code is familiar to us as we have written and executed the same piece of the code in *Chapter 3, Processing Distributed Log Files in Real Time*, except for the statement `flumeStream.flatMap......`

This statement is actually flattening the results and leverages our `Utility` class `ScalaLogAnalyzer` for converting each line of log data received into a `Map` of key/value pairs, where keys are static and values are extracted from the log files by applying the pattern. Finally, the `Map` containing the key/value pair is wrapped into the RDD for further analysis. The values in `Map` should look similar to the following illustration:

```
("method", "GET"),
("request","/twiki/bin/edit/Main/Double_bounce_sender?topicparent=
Main.ConfigurationVariables"),
("size","12846"),
("date","07/Mar/2004:16:05:49-0800"),
("IP","64.242.88.10")
.........
```

Next we have defined a new function `executeTransformations(...)` which accepts the flattened RDD and will define all transformation operations.

Going forward, we will only define the structure of `executeTransformations()` and leave everything else the same.

Now let's enhance our `executeTransformations()` method and add some transformation functions to solve some of the real-world problem statements. We will do this in the form of questions and answers, where we will take up different scenarios and then provide the structure of our function to solve the given problem statement.

- For `EachFunction`: Let's focus on the following problem scenario to understand the implementation of `forEachFunction`:

    Scenario: How do you print all the values of the transformed data?

    Solution: Let's define one more function `printLogValue(...)`, which will use the `forEachRDD` function of `DStream.Scala` for printing all key/values of the transformed log data on the console.

    ```
    def printLogValues(stream:DStream[(String,String)],
    streamCtx: StreamingContext){
      //Implementing ForEach function for printing all the
      data in provided DStream
      stream.foreachRDD(foreachFunc)
    ```

```
//Define the forEachFunction and also print the values
on Console
def foreachFunc = (rdd: RDD[(String,String)]) => {

  //collect() method fetches the data from all
  partitions and "collects" at driver node.
  //So in case data is too huge than driver may crash.
  //In production environments we persist this RDD data
  into HDFS or use the rdd.take(n) method.

  val array = rdd.collect()
  println("---------Start Printing Results----------")
  for(dataMap<-array.array){
    print(dataMap._1,"-----",dataMap._2)
  }
  println("---------Finished Printing Results----------
  ")
  }
}
```

In the preceding implementation, we have used `forEachRDD` to unwrap all the values given in the stream of RDDs and then print the same on the console.

We can then invoke this method from our `executeTransformation(...)` method which should look something like this:

```
/**
 * Define and execute all Transformations to the log
 data
 */
def executeTransformations(dStream:DStream[(String,
String)],streamCtx: StreamingContext){
  //Start - Print all attributes of the Apache Access Log
  printLogValues(dStream,streamCtx)
  //End - Print all attributes of the Apache Access Log
}
```

- `filter(filterFunc)`

    Scenario: How do you count the total number GET requests?

Applying Transformations to Streaming Data

Solution: This is simple if we only have to get the requests whose type is GET and then count them. This can be done by adding the following piece of code to our `executeTransformation` method:

```
dStream.filter(x=> x._1.equals("method") && x._2.contains("GET")).
count().print()
```

The preceding statement filters and retrieves only those elements in the RDD where the value of an element is GET and then further applies the `count()` function and finally prints the total count in the DStream.

- `map(mapFunc)`

    Scenario: How do you count the total number of distinct requests for requested URLs?

    Solution: The solution is again simple. We will filter the DStream and retrieve all the keys containing the "request" as a key and then define a map function to build a new key/value pair of URLs, where key will be the actual URL and the value will be the count. Finally, we will use the `reduceByKey` operation to count all the distinct URLs and print the same on the console.

    ```
    val newStream = dStream.filter(x=>x._1.contains
    ("request")).map(x=>(x._2,1))
    newStream.reduceByKey(_+_).print(100)
    ```

Finally to execute the preceding program, perform the following steps:

1. Export your project as a JAR file and name it as `Spark-Examples.jar` and save this JAR file in the root of `$SPARK_HOME`.

2. Next, open your Linux console and browse to `$SPARK_HOME` and execute the following command to deploy your Spark Streaming job:

    ```
    $SPARK_HOME/bin/spark-submit --class chapter.four.
    ScalaTransformLogEvents --master <SPARK-MASTER-URL> Spark-
    Examples.jar
    ```

As soon as you click on *Enter* and execute the preceding command you should see that all your transformations are working and the results are printed on the console.

Refer to `https://en.wikipedia.org/wiki/Higher-order_function` to know more about high-order functions.

In this section, we have discussed various higher-order and functional operations and their usage in Spark Streaming. Let's move on and discuss transform and windowing operations on streams.

Transform operations

Transform operations, `def transform(....)`, are special types of operation which can be applied to any RDD in DStreams and perform any RDD-to-RDD operations. This method is also used to merge the two Spark worlds, the batch and the streaming one. You can create RDDs using batch processes and merge with RDDs created using Spark Streaming. It even helps in code reusability across Spark batch and streaming, where you may have written functions in your Spark batch applications which we now want to use in our Spark Streaming application.

For example, let's assume that in order to compute the same log data using a Spark batch application you have written a function similar to the following one:

```
val functionCountRequestType = (rdd:RDD[(String,String)]) => {
    rdd.filter(f=>f._1.contains("method"))
.map(x=>(x._2,1))
.reduceByKey(_+_)
}
```

The preceding function takes up an RDD and computes the count of requests (`GET` or `POST`) being served by the server.

Now we will add the preceding function to our `ScalaTransformLogEvents` class and, in the `main` method, we will write something like this:

```
val transformedRDD = dStream.transform(functionCountRequestType)
```

Next, we will invoke the `updateStateByKey` operation to keep a running count of the requests for all the time. `updateStateByKey` creates and returns a new "state" of DStream and the keys defined in the DStream.

The new state of each key is generated by applying the given function on the previous state of the key and the new values of each key. It is also important to enable the "checkpoint" while using `updateStateByKey` so that our streaming application is resilient to failures unrelated to the application logic and the state of the keys at different time intervals are not lost.

Applying Transformations to Streaming Data

Let's move forward and define one more function which defines how the state of a key should be updated, whether it should be adding, deriving average or anything else:

```
val functionTotalCount = (values: Seq[Int], state: Option[Int])=>{
    Option(values.sum + state.sum)
}
```

In the preceding function, the new state is derived by adding the new and the old state of the key.

Finally, we will enable "checkpoint", invoke `updateStateByKey`, and print the values either using `forEach` or the utility function, `print()`:

```
streamCtx.checkpoint("checkpointDir")
transformedRDD.updateStateByKey(functionTotalCount).print(100)
```

In this section we have discussed the `transform` functions which encourage reusability of the code where Spark Streaming jobs can use the functions defined in Spark batch jobs.

Let's move on to the next section and talk about windowing and sliding window operations.

Windowing operations

In real-time streaming, it is important to define the scope and size of data which needs to be analyzed and queried.

Micro-batching provides some flexibility where we can accumulate events and then start our computing jobs but in micro-batching all batches of data are independent and only contain new events as they appear or are received by the streams, they do not add new events to the older events without changing the batch size.

Chapter 4

Windowing functions provide exactly the same functionality where they define the scope of the data which needs to be analyzed, as in the last 10 minutes, 20 minutes, and so on, and further slide this window by a configured interval like one or two minutes.

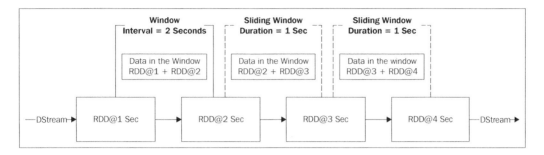

Let's consider the preceding illustration where the interval of the window is two seconds and the sliding window is one second, which means that, at the end of every one second interval, our DStream will contain the data for last two seconds.

DStreams provide a variety of windowed operations. Let's see a few of the windowed operations provided by DStream.

> For further details, you can refer to DStream API https://spark.apache.org/docs/1.3.0/api/scala/index.html#org.apache.spark.streaming.dstream.DStream where all operations which end in "window" provide windowed computations.

Let's extend our `ScalaTransformLogEvents` example and perform the following steps to implement a few of the windowing functions:

1. Open and edit `ScalaTransformLogEvents` and the following new method, which defines the windowing operations:

    ```
    /**
    * Window and Sliding Windows
    */

    def executeWindowingOperations(dStream:DStream[(String,
    String)],streamCtx: StreamingContext){

      //This Provide the Aggregated Count of all response
      Codes
    ```

```
println("Printing count of Response Code using
windowing Operation")
val wStream = dStream.window(Seconds(40),Seconds(20))
val respCodeStream = wStream.filter(x=>x._1.contains
  ("respCode")).map(x=>(x._2,1))
respCodeStream.reduceByKey(_+_).print(100)

//This provide the Aggregated count of all response
Codes by using WIndow operation in Reduce method
println("Printing count of Response Code using
reducebyKeyAndWindow Operation")
val respCodeStream_1 = dStream.filter(x=>x._1.contains
  ("respCode")).map(x=>(x._2,1))
respCodeStream_1.reduceByKeyAndWindow
  ((x:Int,y:Int)=>x+y,Seconds(40),Seconds(20)).
print(100)

//This will apply and print groupByKeyAndWindow in the
Sliding Window
println("Applying and Printing groupByKeyAndWindow in
a Sliding Window")
val respCodeStream_2 = dStream.filter(x=>x._1.contains
  ("respCode")).map(x=>(x._2,1))
respCodeStream_2.groupByKeyAndWindow
(Seconds(40),Seconds(20)).print(100)

}
```

In the preceding function we have defined three variations of the windowing operations:

- `window(WindowDuration, SlideDuration`: This operation works directly on the stream and provides all elements of DStream within the given duration.
- `reduceByKeyAndWindow(reduceFunc, WindowDuration, SlideDuration`: This operation works with the **reduce** function where it applies the **reduce** function only on the RDDs in the sliding window.
- `groupByKeyAndWindow(WindowDuration, SlideDuration)`: Applies the `groupBy` operation on the available keys within the duration as specified by `SlidingDuration`.

> It is important to note that the duration for windows and sliding windows should be multiples of batching intervals defined for DStream. In our case we have defined 10 seconds, so all our window operations should contain multiples of 10.

2. Next, at the end of the `executeTransformation` function, we will invoke this new method and write the following mentioned piece of code:

   ```
   //Start - Windowing Operation
   executeWindowingOperations(dStream,streamCtx)
   //End - Windowing Operation
   ```

And we are done!

In order to execute the preceding code, perform the same steps as in the previous section *Functional operations*.

In this section, we discussed a few of the windows operations exposed by Spark Streaming. Let's move on to the next section where we will talk about the various configurations and parameters available for tuning our Spark application.

Performance tuning

Spark provides various configuration parameters which if used efficiently can significantly improve the overall performance of your Spark Streaming job. Let's look at a few of the features which can help us in tuning our Spark jobs.

Partitioning and parallelism

Spark Streaming jobs collect and buffer data at regular intervals (batch intervals) which is further divided into various stages of execution to form the execution pipeline. Each byte in the dataset is represented by RDD and the execution pipeline is called a **Direct Acyclic Graph** (**DAG**).

The dataset involved in each stage of the execution pipeline is further stored in the data blocks of equal sizes which is nothing more than the partitions represented by the RDD.

Lastly, for each partition, we have exactly one task allocated or executed.

So the parallelism of your job directly depends on the number of partitions configured for your jobs, which can be controlled by defining `spark.default.parallelism` in `$SPARK_HOME/conf/spark-defaults.conf`. It needs to be configured correctly so that you get a sufficient amount of parallelism for your Spark jobs. The general rule is to configure parallelism by at least twice the number of total cores in the cluster but that is a bare minimum value which gives us a starting point and it may vary for different workloads.

Unless specified by the RDDs, Spark by default uses `org.apache.spark.HashPartitioner` and the default value for the maximum number of partitions will be the same as the number of partitions in the largest upstream RDD.

Refer to `http://www.bigsynapse.com/spark-input-output` for more details on understanding partitioning and parallelism.

Serialization

Serialization is another area of focus for performance tuning Spark jobs. Serialization and deserialization are required for not only shuffling data between worker nodes but also when serializing RDDs to disk. By default, Spark utilizes the Java serialization mechanism which is compatible with most file formats but is also slow.

We can switch to Kryo Serialization (`https://github.com/EsotericSoftware/kryo`), which is very compact and faster than Java serialization. Though it does not support all serializable types it is still much faster than the Java serialization mechanism for all the compatible file formats. We can configure our jobs to use `KyroSerializer` by configuring `spark.seralizer` in our `SparkConf` object.

```
conf.set("spark.serializer", "org.apache.spark.serializer.
KryoSerializer")
```

`KyroSerializer` by default stores the full class names with their associated objects in Spark executors memory which again is a waste of memory so, to optimize, it is advisable to register all required classes in advance with `KyroSerializer` so that all objects are mapped to Class IDs and not with full class names. This can be done by defining explicit registrations of all required classes using `SparkConf.registerKryoClasses(….)`.

Refer to the Kyro documentation (`https://github.com/EsotericSoftware/kryo`) for more optimization parameters and compatible file formats.

Spark memory tuning

Spark is a JVM-based execution framework, so tuning the JVMs for the right kind of workloads can also significantly improve the overall response time of our Spark Streaming jobs. To being with, there are a couple of areas where we should focus.

Garbage collection

As a first step, we need uncover the current GC behavior and statistics and, for that, we should configure following parameters in `$SPARK_HOME/conf/spark-defaults.conf`:

```
spark.executor.extraJavaOptions = -XX:+PrintFlagsFinal
-XX:+PrintReferenceGC -Xloggc:$JAVA_HOME/jvm.log -XX:+PrintGCDetails
-XX:+PrintGCTimeStamps -XX:+PrintAdaptiveSizePolicy
```

Now your GC details are printed and available in `$JAVA_HOME/jvm.log`. We can further analyze the behavior of our JVM and then apply optimization techniques.

Refer to https://databricks.com/blog/2015/05/28/tuning-java-garbage-collection-for-spark-applications.html for more details on various optimization techniques and tuning GC for Spark applications.

Object sizes

Optimizing the size of the objects which are stored in the memory can also improve the overall performance of your application. Here are a few tips which can help us to improve the memory consumed by our objects:

1. Avoid using wrapper objects or pointer bases, data structures, or nested data structures with a lot of small objects.
2. Use arrays of objects or primitive types in your data structures. You can also use the `fastutil` (http://fastutil.di.unimi.it/) library which provides faster and optimized collection classes.
3. Avoid strings or custom objects; instead use numeric IDs for the objects.

Executor memory and caching RDDs

Another option is to configure the memory of the Spark executors and also decide the appropriate size given in the memory to our Spark jobs for caching the RDDs.

Executor memory can be configured by defining the `spark.executor.memory` property in the `SparkConf` object or `$SPARK_HOME/conf/spark-defaults.conf` or we can also define it while submitting our jobs:

```
val conf = new SparkConf().set("spark.executor.memory", "1g")
```

Or:

```
$SPARK_HOME/bin/spark-submit --executor-memory 1g …..
```

The Spark framework, by default, takes up 60 percent of the executor's configured memory for caching RDDs, which leaves only 40 percent as available memory for the execution of your Spark jobs. This may not be sufficient and if you see full GCs or slowness in your tasks or may be experiencing out of memory, then you can reduce the cache size by configuring `spark.storage.memoryFraction` in your `SparkConf` object:

```
val conf = new SparkConf().set("spark.storage.memoryFraction",
  "0.4")
```

The preceding statement reduces the memory allocated for caching RDDs to 40 percent.

Finally, we can also think about using off-heap caching solutions which do not use a JVM like Tachyon (`http://tachyon-project.org/Running-Spark-on-Tachyon.html`).

> For more details on performance features, refer to `https://spark.apache.org/docs/1.3.0/tuning.html` and `https://spark.apache.org/docs/1.3.0/configuration.html` for various available configuration parameters.

In this section we have talked about various aspects of tuning our Spark jobs but no matter how much we discuss it, performance is always tricky and there will always be new discoveries which may require expert advice. So, for all that expert advice, post your queries to the Spark Community (`https://spark.apache.org/community.html`).

Summary

In this chapter, we have discussed in detail the various available transformations, starting from aggregations and also including advance functions like windowing and sliding windows. We also discussed various aspects of Spark Streaming which should be considered while performance tuning our Spark Streaming jobs.

In the next chapter, we will discuss persisting log analysis data and integration with various other NoSQL databases.

5
Persisting Log Analysis Data

Systems like Hadoop and Spark have reduced the overall cost of solutions and infrastructure. They have revolutionized the industry with low cost but robust and stable frameworks which are scalable/extendable/customizable and can process a variety of data formats.

There are a bunch of enterprise products for performing **ETL (Extraction, Transformation**, and **Loading**) operations but, when we have to deal with the huge amount of data or varied data sources or produce reports in an hour, these tools are not useful. As a result, architects and developers have been evaluating and implementing Hadoop and Spark-like systems just for ETL use cases where the source or raw data is processed or transformed and finally stored in external systems like Oracle or Teradata or simply generate the reports to dump on the filesystem.

The benefit of using Hadoop or Spark as an ETL framework is that it provides a low cost solution where all kinds of data processing is performed on low cost and commodity hardware and eventually reduces the overall size of raw data from TBs to GBs/MBs and finally stores the processed data back into Oracle or Teradata for further analysis. Hadoop or Spark-like frameworks can be scaled out vertically, and adding nodes to the existing cluster can be done in no time without bringing down the whole cluster and also providing no **SPOF (single point of failure**).

The real advantage here is that systems like Oracle or Teradata have to store only transformed data which is less than 50 percent of the overall size of the raw data, which eventually saves on the overall cost of the hardware and software licenses for Oracle and Teradata.

We talked about extraction and transformation in previous chapters. In this chapter we will discuss the operations provided by Spark for writing data (loading) to external systems. We will also talk about the integration of Spark with various popular NoSQL databases.

This chapter will cover the following points:

- Output operation in Spark Streaming
- Integration with NoSQL DB – Cassandra

Output operations in Spark Streaming

Spark Streaming provides various standard and customizable output operations over DStreams which can read the data from RDDs and can either save it to Hadoop or a text file, or print on the console. Let's see these output operations provided by Spark Streaming:

- `print()`: This is the one of the basic functions for printing elements of RDDs stored within DStreams. It executes on the driver node and, by default, prints the first 10 elements of every batch of data in DStreams on the driver console. There is also an overloaded function `print(numElements:Int)` which provides the flexibility to print more than 10 elements.

- `saveAsHadoopFiles(...)`: This operation provides the integration with the Hadoop/HDFS where it persist the elements of the DStream in **HDFS (Hadoop Distributed File System)**. This method works with the old MapReduce API which was available with the Hadoop = <0.20 version and can be invoked only on the DStreams containing a key/value pair. It is exposed by `org.apache.spark.streaming.dstream.PairDStreamFunctions.scala`, so it is not available in the standard DStreams.

- `saveAsNewAPIHadoopFiles(...)`: Similar to `saveAsHadoopFiles(...)` with the difference that it leverages a new MapReduce API which is available with the Hadoop > 0.20 version.
- `saveAsTextFiles(...)`: Persists the elements of the invoking DStream as a text file on the local filesystem or any other provided location.
- `saveAsObjectFiles(...)`: Persists the elements of the invoking DStream as a sequence file at the provided location on the HDFS/filesystem.
- `forEachRDD(..)`: `forEachRDD` is a powerful, generic and special type of operation which can be extended and customized for integration and persisting data in any of the external systems which may not available as part of the standard Spark distribution, for example, persisting data in RDBMS, submitting data to REST services or MQ systems, and many more. It is similar to `forEach()`, where it takes up an arbitrary function as an argument which is further applied over each RDD in DStreams. This output operation is executed on RDDs which are available over the worker nodes. For example, let's assume that we need to submit the elements of each RDD into a messaging queue, so our implementation would look something like this:

```
dStream.forEachRDD{ //Line-1
//assuming that this method provides the connection to
underlying MQ infrastructure
val conn = getConnection(…) // Line-2
   rdd=> rdd.forEach{ // Line-3
   //Function to create messages and post to Queues/
   //Topics
   createTextMessages().post(conn) // Line-4
   }
}
```

Persisting Log Analysis Data

Although logically the preceding code will work and we will be able to post data to queues and queues/topics, the preceding methods have overheads and sometimes may produce exceptions because of the following reasons:

- The getConnection() function in the preceding code is executed over the driver while the rest of the function will be executed over the worker nodes, which in turn means that the connection object needs to be serialized and will be sent to the worker nodes, which is not a good idea.

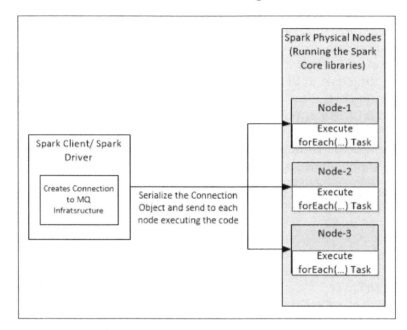

The preceding illustration shows the process of creating a connection at driver nodes and then serializing the connection object and sending it to physical or worker nodes.

A better approach is to create connection on the worker nodes themselves and move the code for creating the connection just before Line-4.

- Assuming that we created a connection just before Line-4, again this is not the best approach as now we are creating a connection for each record in the RDD. So, in order to optimize it, we can implement forEachPartition() and then create connections for each partition, instead of each record. Our final implementation would look something like this:

```
dStream.forEachRDD{ //Line-1
  rdd=> forEachPartition{ //Line-2
  val conn = getConnection(...) // Line-3
```

```
partition=> partition.forEach(record =>
createTextMessages().post(conn)) //Line-4
  }
}
```

The following illustration shows the best way to create a connection to external systems like MQ for each partition:

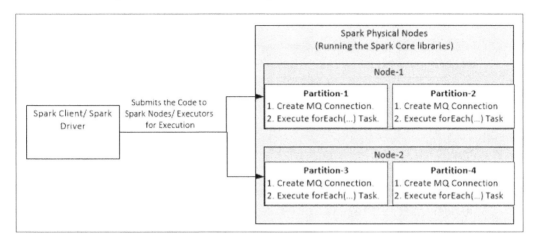

 Java API for DStreams, `org.apache.spark.streaming.api.java.JavaDStream`, can also be used to invoke all of the preceding output operations.

Let's see a working piece of code which will leverage the output operations and persists data in filesystems and Hadoop.

Perform the following steps to set up Hadoop and HDFS which we will be using for persisting our log file data:

1. Download Hadoop 2.4.0 distribution from https://archive.apache.org/dist/hadoop/common/hadoop-2.4.0/hadoop-2.4.0.tar.gz and extract the archive to any folder of your choice on the same machine where you configured Spark.

2. Open the Linux shell and execute:
   ```
   export HADOOP_PREFIX=<path of your directory where we extracted Hadoop>
   ```

3. Follow the steps defined at http://hadoop.apache.org/docs/r2.5.2/hadoop-project-dist/hadoop-common/SingleCluster.html for single node setup. After completing the "prerequisites" defined in the given link, you can follow the setup instructions defined either for "pseudo-distributed mode" or "fully-distributed mode". For our needs "pseudo-distributed mode" will work but that does not stop you from trying the latter one.

4. Once you have completed the setup, open your Linux shell and execute the following commands:

   `$HADOOP_PREFIX/bin/hdfs namenode -format`

   `$HADOOP_PREFIX/sbin/start-dfs.sh`

5. The first command will format `namenode` and make the filesystem ready for use and with the second command we are starting the minimum required Hadoop services which will include namenode and secondary namenode.

6. Next, let's execute these commands to create a directory structure in HDFS where we will store our data:

   `$HADOOP_PREFIX/bin/hdfs dfs -mkdir /spark`

   `$HADOOP_PREFIX/bin/hdfs dfs -mkdir /spark/streaming`

   `$HADOOP_PREFIX/bin/hdfs dfs -mkdir /spark/streaming/oldApi`

   `$HADOOP_PREFIX/bin/hdfs dfs -mkdir /spark/streaming/newApi`

   `$HADOOP_PREFIX/bin/hdfs dfs -mkdir /spark/streaming/sequenceFiles`

If everything goes right and there are no exceptions then open your browser and browse to http://localhost:50070/explorer.html#/ and you will be able to see the empty directories created by the preceding commands, as shown in the following illustration:

Browse Directory

/spark/streaming

Permission	Owner	Group	Size	Replication	Block Size	Name
drwxr-xr-x	ec2-user	supergroup	0 B	0	0 B	newApi
drwxr-xr-x	ec2-user	supergroup	0 B	0	0 B	oldApi
drwxr-xr-x	ec2-user	supergroup	0 B	0	0 B	sequenceFiles

The preceding image shows the HDFS filesystem explorer where we can browse, view and download any of the files created by the users using HDFS APIs.

Chapter 5

> Refer to http://hadoop.apache.org/ for more information on Hadoop and HDFS.

Perform the following steps to implement output operations over DStreams and persist data in the HDFS and local filesystem:

1. Open your Linux shell and create a directory by executing the following command:

   `mkdir $SPARK_HOME/outputDir`

2. Next, we will extend our Spark-Examples project and create a package called chapter.five and a Scala object ScalaPersistLogData and add the following piece of code:

   ```
   package chapter.five

   import org.apache.spark.SparkConf
   import org.apache.spark.streaming._
   import org.apache.spark.streaming.flume._
   import org.apache.spark.storage.StorageLevel
   import org.apache.spark.rdd._
   import org.apache.spark.streaming.dstream._
   import java.net.InetSocketAddress
   import java.io.ObjectOutputStream
   import java.io.ObjectOutput
   import java.io.ByteArrayOutputStream
   import org.apache.spark.SparkContext
   import chapter.four.ScalaLogAnalyzer
   import org.apache.hadoop.io._
   import org.apache.hadoop.mapreduce._
   import org.apache.hadoop.mapred.JobConf
   import org.apache.hadoop._

   object ScalaPersistLogData {

     def main(args:Array[String]){

       /** Start Common piece of code for all kinds of Output
       Operations*/
       println("Creating Spark Configuration")
       val conf = new SparkConf()
   ```

[105]

```
        conf.setAppName("Apache Log Persister")
        println("Retreiving Streaming Context from Spark Conf")
        val streamCtx = new StreamingContext(conf, Seconds(10))

        var addresses = new Array[InetSocketAddress](1);
        addresses(0) = new InetSocketAddress("localhost",4949)

        val flumeStream = FlumeUtils.createPollingStream
        (streamCtx,addresses,StorageLevel.
        MEMORY_AND_DISK_SER_2,1000,1)

        //Utility class for Transforming Log Data
        val transformLog = new ScalaLogAnalyzer()
        //Invoking Flatmap operation for flattening the results
        //and convert them into Key/Value pairs
        val newDstream = flumeStream.flatMap { x =>
        transformLog.tansformLogData(new
        String(x.event.getBody().array())) }

        /** End Common piece of code for all kinds of Output
        Operations*/

        /**Start - Output Operations */
        persistsDstreams(newDstream,streamCtx)
        /**End - Output Operations */

        streamCtx.start();
        streamCtx.awaitTermination();
    }
```

The main method of our new class, `ScalaPersistLogData`, remains almost the same, where we have defined the common piece of code which consumes the data from Spark Sink. Refer to the *Installing and configuring Flume* section of *Chapter 3, Processing Distributed Log Files in Real Time*, for setting up Spark Sink.

But at the end of the `main()` method we have invoked a new method `persistsDstreams(...)` which will contain all our output operations.

3. Next, we will define a new method `persistsDstreams(...)` and add the following piece of code to `ScalaPersistLogData.scala`:

```
/**
    * Define and execute all Output Operations over
    DStreams
```

```
*/
def persistsDstreams(dStream:DStream[(String,
String)],streamCtx: StreamingContext){

    //Writing Data as Text Files on Local File system.
    //This method takes 2 arguments: -
    //1."prefix" of file, which would be appended with Time
    //(in milliseconds) by Spark API's
    //2."suffix" of the file
    //The final format will be
    //"<prefix><Milliseconds><suffix>"
    dStream.saveAsTextFiles("/home/ec2-user/softwares/spark-
    1.3.0-bin-hadoop2.4/outputDir/data-", "")

    //Creating an Object of Hadoop Config with default Values
    val hConf = new JobConf(new
    org.apache.hadoop.conf.Configuration())

    //Defining the TextOutputFormat using old API's
    //available with =<0.20
    val oldClassOutput = classOf[org.apache.hadoop.mapred.
    TextOutputFormat[Text,Text]]
    //Invoking Output operation to save data in HDFS using
    //old API's
    //This method accepts following Parameters: -
    //1."prefix" of file, which would be appended with Time
    //(in milliseconds) by Spark API's
    //2."suffix" of the file
    //3.Key - Class which can work with the Key
    //4.Value - Class which can work with the Key
    //5.OutputFormat - Class needed for writing the Output
    //in a specific Format
    //6.HadoopConfig - Object of Hadoop Config
    dStream.saveAsHadoopFiles("hdfs://localhost:9000/
    spark/streaming/oldApi/data-", "", classOf[Text],
    classOf[Text], oldClassOutput ,hConf )

    //Defining the TextOutputFormat using new API's
    available with >0.20
    val newTextOutputFormat = classOf[org.apache.hadoop.
    mapreduce.lib.output.TextOutputFormat[Text, Text]]

    //Invoking Output operation to save data in HDFS using
    new API's
```

Persisting Log Analysis Data

```
//This method accepts same set of parameters as
"saveAsHadoopFiles"
dStream.saveAsNewAPIHadoopFiles
("hdfs://localhost:9000/spark/streaming/
newApi/data-", "", classOf[Text], classOf[Text],
newTextOutputFormat ,hConf )

//Defining saveAsObject for saving data in form of
//Hadoop Sequence Files
dStream.saveAsObjectFiles
("hdfs://localhost:9000/spark/
streaming/sequenceFiles/data-")

//Using forEachRDD for printing the data for each Partition
dStream.foreachRDD(
    rdd => rdd.foreachPartition(
    data=> data.foreach(
        //Printing the Values which can be replaced by
        //custom code
        //for storing data in any other external
        //System.
        tup => System.out.println("Key = "+tup._1+",
        Value = "+tup._2)
        )
      )
    )
}
```

And we are done! Now let's execute this preceding piece of code and see the output in the HDFS and filesystem.

Perform the following steps to execute the preceding piece of code:

1. Assuming that your Spark job from *Chapter 4, Applying Transformations to Streaming Data* is still running, stop it by pressing *Ctrl* + *C*.
2. Next, export your project as a JAR file, name it `Spark-Examples.jar` and save it in the root of `$SPARK_HOME`.
3. Next, open your Linux console and navigate to `$SPARK_HOME` and execute the following command to deploy your Spark Streaming job:

   `$SPARK_HOME/bin/spark-submit --class chapter.five. ScalaPersistLogData --master <SPARK-MASTER-URL> Spark-Examples.jar`

As soon as you execute the preceding command and hit *Enter*, you will see that all your transformed data is persisted into different directories in HDFS and the filesystem which should look something similar to the following illustration:

The preceding illustration shows the directories and files created in HDFS by the `saveAsNewAPIHadoopFile(...)` method. You can further click on any of the directories and open the contents of the files.

The preceding illustration shows the directories and files created in HDFS by the `saveAsHadoopFile(...)` method. You can further click on any of the directories and open the contents of the files created in the directories.

The same structure would be created by `saveAsObjectFiles(...)` in HDFS in the `sequenceFiles` folder and `saveAsTextFiles(...)` in `$SPARK_HOME/outputDir/`.

Apart from the preceding illustration you will also see the data printed on the console and that is done by the `forEachRDD(func)` method.

In this section we have discussed various output operations exposed by the DStreams API. We also configured and set up Hadoop/HDFS and finally implemented output operations for storing log data in the various directories on the filesystem and HDFS.

Let's move on to the section where we will talk about Spark integration with NoSQL systems like Cassandra for storing and retrieving our log data.

Integration with Cassandra

Apache Cassandra, http://cassandra.apache.org/, is a massively distributed database for handling large data across data centers. It is a linearly scalable NoSQL (non-relational) open source database which offers high availability with ease of operation. It also offers a low cost solution for commodity hardware or cloud infrastructure but with proven fault tolerance.

The integration of a NoSQL database like Cassandra with Spark Streaming not only provides the flexibility to downstream systems like web applications, portals or mobile apps for consumption of the processed data according to convenience or requirements, but can also be used as the data source for Spark for further deep analytics.

DataStax, www.datastax.com, the company which offers a commercial product DataStax Enterprise (built on top of Apache Cassandra) realized the opportunity and provided an excellent open source driver/API for integration between Spark and Cassandra. The Spark-Cassandra driver can be used in our Spark applications where we can directly load the data from Cassandra and use it our Spark applications for further analysis. This driver also provides server-side filters which only load selected or relevant datasets so that our Spark application only works with the portion of data of interest, saving resources (network, memory, CPU). This driver provides all APIs for performing CRUD operations over Cassandra and it also provides specific operations on DStreams so that you do not have to define forEach(...), for example, it defines DStream.saveToCassandra(...) for storing the data of each RDD in the Cassandra tables.

Let's extend our Spark project, Spark-Examples, and write Spark Streaming jobs for consuming streaming web logs and then further persisting the same data in Cassandra.

Installing and configuring Apache Cassandra

Perform the following steps to install and configure Apache Cassandra:

1. Download and extract Apache Cassandra 2.1.7 from http://www.apache.org/dyn/closer.lua/cassandra/2.1.7/apache-cassandra-2.1.7-bin.tar.gz on the same machine where we installed our Spark and Flume software.

2. Execute the following command on your Linux console and define the environment variable CASSANDRA_HOME as the environment variable which will point to the directory where we have extracted the downloaded archive file.:

   `export CASSANDRA_HOME = <location of the Archive>`

3. Next, on the same console, execute the following command to bring up your Cassandra database with the default configuration:

   `$CASSANDRA_HOME/bin/cassandra`

   The preceding command will bring up your Cassandra database which is now ready to serve the user request but, before that, let's create a keyspace and tables where we will store our data.

4. Execute the following command to open the Cassandra **CQL** (**Cassandra Query Language**) console:

   `$CASSANDRA_HOME/bin/cqlsh`

   CQLSH is the command line utility which provides SQL-like syntax to perform CRUD operations on Cassandra databases.

5. Next, execute the following CQL commands on your CQLSH to create a keyspace and table in your Cassandra database:

   `CREATE KEYSPACE logdata WITH replication = {'class': 'SimpleStrategy', 'replication_factor': 1 };`

   `CREATE TABLE logdata.apachelogdata(ip text PRIMARY KEY, client text,user text,date text,method text,request text,protocol text,respcode text,size text);`

   `select * from logdata.apachelogdata;`

   The following illustration shows the output of the CQL commands which creates the keyspace `logdata` and table `apachelogdata` within the keyspace and also executes the `select` query:

```
sumit@localhost $ $CASSANDRA_HOME/bin/cqlsh
Connected to Test Cluster at 127.0.0.1:9042.
[cqlsh 5.0.1 | Cassandra 2.1.7 | CQL spec 3.2.0 | Native protocol v3]
Use HELP for help.
cqlsh> CREATE KEYSPACE logdata WITH replication = {'class': 'SimpleStrategy', 'replication_factor': 1 };
cqlsh> CREATE TABLE logdata.apachelogdata(ip text PRIMARY KEY, client text,user text,date text,method text,request text,protocol text,respcode text,size text);
cqlsh> select * from logdata.apachelogdata;

 ip | client | date | method | protocol | request | respcode | size | user
----+--------+------+--------+----------+---------+----------+------+------

(0 rows)
cqlsh>
```

Our Cassandra setup is complete and now we will code and configure our Spark application to insert data in the `logdata.apachelogdata` table.

Configuring Spark for integration with Cassandra

Perform the following steps for configuring and integrating Spark platform with Cassandra:

1. Download the following JAR files and store them in the `$CASSANDRA_HOME/lib/`.

    - Spark-Cassandra connector: http://central.maven.org/maven2/com/datastax/spark/spark-cassandra-connector_2.10/1.3.0-M1/spark-cassandra-connector_2.10-1.3.0-M1.jar
    - Cassandra Core driver: http://search.maven.org/remotecontent?filepath=com/datastax/cassandra/cassandra-driver-core/2.1.5/cassandra-driver-core-2.1.5.jar
    - Spark-Cassandra Java library: http://search.maven.org/remotecontent?filepath=com/datastax/spark/spark-cassandra-connector-java_2.10/1.3.0-M1/spark-cassandra-connector-java_2.10-1.3.0-M1.jar
    - Other libraries:
        - http://central.maven.org/maven2/org/joda/joda-convert/1.2/joda-convert-1.2.jar
        - http://central.maven.org/maven2/joda-time/joda-time/2.3/joda-time-2.3.jar
        - http://central.maven.org/maven2/com/twitter/jsr166e/1.1.0/jsr166e-1.1.0.jar

2. Next, the your `$SPARK_HOME/conf/spark-default.conf` file and append the path given for `spark.executor.extraClassPath` and `spark.driver.extraClassPath` with the physical path of the libraries downloaded in the previous step.

3. Next, append and add the dependencies of the following Cassandra library for the `spark.executor.extraClassPath` and `spark.driver.extraClassPath` variables:

    ```
    $CASSANDRA_HOME/lib/apache-cassandra-2.1.7.jar
    $CASSANDRA_HOME/lib/apache-cassandra-clientutil-2.1.7.jar
    $CASSANDRA_HOME/lib/apache-cassandra-thrift-2.1.7.jar
    ```

```
$CASSANDRA_HOME/lib/cassandra-driver-internal-only-2.5.1.zip
$CASSANDRA_HOME/lib/thrift-server-0.3.7.jar:
$CASSANDRA_HOME/lib/guava-16.0.jar
```

4. Next, save the `spark-default.conf` file and restart Spark master and worker to check everything works as expected without any exceptions or errors.

We are done with the configurations. Let's now write the Spark Streaming code for capturing log analysis data to further store in Cassandra.

Coding Spark jobs for persisting streaming web logs in Cassandra

Let's extend our `Spark-Examples` project and perform the following steps to create a Spark job for persisting web logs in Cassandra:

1. Open the `Spark-Examples` project and edit `chapter.four.ScalaLogAnalyzer` and add the following piece of code just before the closing bracket of the class `chapter.four.ScalaLogAnalyzer`:

```
def tansformLogDataIntoSeq(logLine: String):Seq[(String,String,String,String,String,String,String,String,String)] ={
    //Pattern which will extract the relevant data from
    Apache Access Log Files """^(\S+) (\S+) (\S+)
    \[([\w:/]+\s[+\-]\d{4})\] "(\S+) (\S+) (\S+)"
    (\d{3}) (\S+)""";
    val PATTERN = Pattern.compile(LOG_ENTRY_PATTERN);
    val matcher = PATTERN.matcher(logLine);

    //Matching the pattern for the each line of the Apache
    access Log file
    if (!matcher.find()) {
      System.out.println("Cannot parse logline" + logLine);
    }
    //Finally create a Key/Value pair of extracted data and
    //return to calling program
    createSeq(matcher);

}
```

Persisting Log Analysis Data

```
def createSeq(m:Matcher):Seq[(String,String,String,
String,String,String,String,String,String)] = {
    Seq((m.group(1),m.group(2), m.group(3), m.group(4),
    m.group(5), m.group(6), m.group(7) , m.group(8),
    m.group(9)))
}
```

The preceding piece of code defines a utility function for converting log log data into iterable collections which we will further use and insert in Cassandra.

> For more information on Seq, refer to http://www.scala-lang.org/api/2.11.5/index.html#scala.collection.Seq.

2. Within the package chapter.five, create a new Scala object ScalaPersistInCassandra and add the following piece of code:

```
package chapter.five

import org.apache.spark.SparkConf
import org.apache.spark.SparkContext
import org.apache.spark.streaming.StreamingContext
import java.net.InetSocketAddress
import org.apache.spark.streaming.flume.FlumeUtils
import org.apache.spark.streaming.Seconds
import org.apache.spark.storage.StorageLevel
import com.datastax.spark.connector.streaming._
import com.datastax.spark.connector._
import chapter.four.ScalaLogAnalyzer
import com.datastax.spark.connector.SomeColumns

object ScalaPersistInCassandra {

  def main(args:Array[String]){

    /** Start Common piece of code for all kinds of Output
    Operations*/
    println("Creating Spark Configuration")
    val conf = new SparkConf()
```

[114]

```
conf.setAppName("Apache Log Persister in Cassandra")
//Cassandra Host Name
println("Setting Cassandra Host Name for getting
Connection")
conf.set("spark.cassandra.connection.host",
"localhost")

println("Retreiving Streaming Context from Spark Conf")
val streamCtx = new StreamingContext(conf, Seconds(10))
var addresses = new Array[InetSocketAddress](1);
addresses(0) = new InetSocketAddress("localhost",4949)
val flumeStream = FlumeUtils.createPollingStream
(streamCtx,addresses,StorageLevel.
MEMORY_AND_DISK_SER_2,1000,1)
//Utility class for Transforming Log Data
val transformLog = new ScalaLogAnalyzer()
//Invoking Flatmap operation to flattening the results
and convert them into Key/Value pairs
val newDstream = flumeStream.flatMap { x =>
transformLog.tansformLogDataIntoSeq(new
String(x.event.getBody().array())) }
/** End Common piece of code for all kinds of Output
Operations*/

//Define Keyspace
val keyspaceName ="logdata"
//Define Table
val csTableName="apachelogdata"
//Invoke saveToCassandra to persist DStream to
Cassandra CF
newDstream
.saveToCassandra(keyspaceName, csTableName,
SomeColumns("ip","client","user","date",
"method","request","protocol","respcode","size"))

streamCtx.start();
streamCtx.awaitTermination();

    }
}
```

Persisting Log Analysis Data

And we are done! Now perform the following steps to execute the preceding piece of code and see the output in Cassandra.

1. Assuming that your Spark cluster is down, start the Spark master, slave and Flume agent. If required, also run your log simulator so that your agent receives the log data.

2. Next, export your project as a JAR file, name it `Spark-Examples.jar` and save it JAR file in the root of `$SPARK_HOME`.

3. Next, open the Linux console and browse to `$SPARK_HOME` and execute the following command to deploy your Spark Streaming job:

   `$SPARK_HOME/bin/spark-submit --class chapter.five.ScalaPersistInCassandra --master <SPARK-MASTER-URL> Spark-Examples.jar.`

4. As soon as the preceding statement has executed, hit *Enter* and your Apache log data is persisted in Cassandra which can be further queried by executing the following command on CQLSH:

   `select * from LogData.ApacheLogData;`

   The preceding illustration shows the output of the `select` command on the CQLSH:

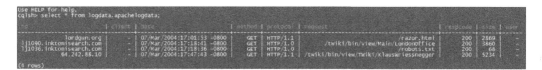

Let's perform the following steps to read and filter data from the Cassandra table based on certain criteria in our Spark Streaming job. We also print the top 10 `GET` requests based on the write time to the Cassandra table.

1. Open and edit `ScalaPersistInCassandra.scala` and add a new method by the name of `readAndPrintData`, shown as follows:

   `def readAndPrintData(streamCtx: StreamingContext){}`

2. Next, invoke this method from the main method just after the initialization of `StreamingContext`:

   ```
   println("Retreiving Streaming Context from Spark Conf")
   val streamCtx = new StreamingContext(conf, Seconds(10))
   //First read the existing data from Cassandra
   readAndPrintData(streamCtx)
   ```

3. Now, edit `readAndPrintData(streamCtx)` and add the following piece of code for printing a specific column of the Cassandra table and also printing all rows:

```
//Reading data from Cassandra and Printing on Console
println("Start - Printing the data from Cassandra.. ")
println("Start - Print All IP's .................. ")
//Prints the first Column (IP) of the table
//Get the reference of the apachelogdata table from the Context
//which further returns the Object of CassandraTableScanRDD
val csRDD = streamCtx.cassandraTable("logdata",
"apachelogdata").collect()
//Now using forEach print only the ip column using the
getString() method
csRDD.foreach ( x => println("IP = " + x.getString("ip")))
println("End - Print All IP's ...................... ")
println("Start - Print All Rows ..................... ")
//Use the Same RDD and print complete Rows just by
//invoking toString() method
csRDD.foreach ( x => println("Cassandra Row = " +
x.toString()))
println("End - Print All Rows ...................... ")
println("End - Printing the data from Cassandra...... ")
```

The preceding piece of code is executed before your job is launched and will print the data on your console or driver where your job is running.

4. Next we will see the usage of server-side filters but, before that, let's execute the following set of commands on the Linux console to open CQLSH and create an index on the column `method`.

   $CASSANDRA_HOME/bin/cqlsh

   CREATE INDEX method_name ON logdata.apachelogdata (method);

   In Cassandra, server-side filters only work with indexed columns or we have to include extra parameters (ALLOW FILTERING) at the end of the CQL query. The advantage of using server-side filters is that the database performs the filtering and provides the relevant data which is more efficient than filtering data at the Spark cluster. In the next steps, we will use the column `method` as our filtering criteria.

5. Edit the `readAndPrintData(streamCtx)` method and just before the closing braces add the following piece of code:

```
println("Start - Print only Filetered Rows ......... ")
//Get the RDD and select the column to be printed and use
//where clause
//to specify the condition.
//Here we are selecting only "ip" column where "method=GET"
val csFilterRDD = streamCtx.cassandraTable("logdata",
"apachelogdata").select("ip").where("method=?","GET")
//Finally print the ip column by using foreach loop.
csFilterRDD.collect().foreach( x => println("IP = " +
x.getString("ip")))
println("End - Print only Filetered Rows ......... ")
```

6. Next, edit the `readAndPrintData(streamCtx)` method and just before the closing braces add the following piece of code for printing the top 10 GET requests based on the "write" time to the Cassandra table:

```
println("Start - Print Top 10 GET request")
//we are using the *writetime* method of CQL which gives
//time(microseconds) of record written in Cassandra
 val csTimeRDD = streamCtx.cassandraTable("logdata",
"apachelogdata").
select("ip","method","date",
"method".writeTime.as("time")).where("method=?","GET")
csTimeRDD.collect().
sortBy(x => calculateDate(x.getLong("time"))).
reverse.take(10).foreach(
     x =>
println(x.getString("ip") + " - " + x.getString("date")+" - "+
x.getString("method")+" - "+calculateDate(x.getLong("time")) ))
println("End - Print Top 10 Latest request ")
```

> The preceding piece of code needs to be used with caution. The ideal way to use it in a production environment with large datasets would be to use server-side filters and perform sorting and ordering in Cassandra itself.

We also need to add a convenient method outside `readAndPrintData(...)` which will convert microseconds into `java.util.Date`:

```
import java.util.Date
import java.util.Calendar

  /**
   * Converting Microseconds to Date
   */
  def calculateDate(data:Long): Date = {
    val cal = Calendar.getInstance
    cal.setTimeInMillis(data/1000)
    cal.getTime
  }
```

The preceding method not only prints the top 10 GET requests based on the "write" time to the Cassandra table but it also helps us in knowing the latency between the records produced by the webserver and the time when it was written in the Cassandra server, which we can easily get by computing the difference between the "write" time and the "date" in the Cassandra table.

> In the production environment it is advised to separate reads from writes, where writes can be streamed in separate job and reads can be scheduled as a recurring Spark job executed after a certain time interval, say every 10 seconds, so that the user can see the top 10 records being refreshed and updated after a certain interval.

We are done with all our changes and now let's compile and run the Spark job again to see that, as per expectations, data from the Cassandra table is printed on the console.

> Refer to the official documentation https://github.com/datastax/spark-cassandra-connector/ for more information on the available APIs and functionality supported by the Spark-Cassandra connector.

In this section we have discussed the integration of Spark and Cassandra. We also talked about a few of the features like server-side filters exposed by the connector.

Summary

In this chapter, we have discussed in detail various output operations. We have further extended our output operation and shown fully fledged integration with Apache Cassandra using the Spark-Cassandra connector.

In the next chapter, we will discuss the integration of Spark Streaming with other advance Spark libraries.

6
Integration with Advanced Spark Libraries

No single software in today's world can fulfill the varied, versatile, and complex demands or needs of enterprises and to be honest neither should it!

Software is made to fulfill specific needs arising out of the enterprises at a particular point in time which may change in future due to many other factors. These factors may or may not be controlled like government policies, business/market dynamics, and many more.

Considering all these factors, integration and interoperability of any software system with internal/external systems/software is pivotal in fulfilling the enterprise needs. Integration and interoperability are categorized as non-functional requirements, which are always implicit and may or may not be explicitly stated by the end users.

Over the period of time, architects have realized the importance of these implicit requirements in modern enterprises and now all the enterprise architectures provide due diligence and provisions in the fulfillment of these requirements. Even the enterprise architecture frameworks such as **The Open Group Architecture Framework (TOGAF)** defines the specific set of procedures and guidelines for defining and establishing the interoperability and integration requirements of modern enterprises.

Spark community realized the importance of both these factors and provided a versatile and scalable framework with certain hooks for integration and interoperability with the different systems/libraries for example data consumed and processed via Spark streams can also be loaded into the structured (table: rows/columns) format and can be further queried using SQL. Even the data can be stored in form of Hive tables in HDFS as persistent tables that will exist even after our Spark program has restarted.

In this chapter, we will discuss about the integration of Spark Streaming with various other advanced Spark libraries such as Spark SQL and Spark GraphX.

This chapter will cover the following points:

- Querying streaming data in real time: Spark SQL
- Graph analysis: Spark GraphX

Querying streaming data in real time

Spark Streaming is developed on the principle of integration and interoperability where it doesn't only provide a framework for consuming data in near real time from varied data sources but at the same time, it also provides the integration with Spark SQL where existing DStreams can be converted into structured data format for querying using standard SQL constructs.

There are many such use cases where SQL on streaming data is a much needed feature, for example, in our distributed log analysis use case, we may need to combine the precomputed datasets with the streaming data for performing exploratory analysis using interactive SQL queries, which is difficult to implement only with streaming operators as they are not designed for introducing new datasets and perform ad hoc queries.

Moreover, SQL's success at expressing complex data transformations derives from the fact that it is based on a set of very powerful data processing primitives that do filtering, merging, correlation, and aggregation, which is not available in the low-level programming languages such as Java/C++ and may result in long development cycles and high maintenance costs.

Let's move forward and first understand few things about Spark SQL and then, we will also see the process of converting the existing DStreams into the structured formats.

Understanding Spark SQL

Spark SQL is one of the modules developed over Spark framework for processing structured data, which is stored in the form of rows and columns. At a very high level, it is similar to the data residing in RDBMS in form rows and columns and then SQL queries are executed for performing analysis, but Spark SQL is much more versatile and flexible as compared to RDBMS. Spark SQL provides distributed processing of SQL queries and can be compared to frameworks like Hive, Impala, or Drill. Here are a few notable features of Spark SQL:

- It is capable of loading data from a variety of data sources, such as text files, JSON, Hive, HDFS, Parquet format and of course RDBMS too, so that we can consume, join and process datasets from different and varied data sources
- It supports static and dynamic schema definition for the data loaded from various sources, which helps in defining schema for known data structures/types and also for those datasets where the columns and their types are not known until runtime
- It can work as a distributed query engine using thrift JDBC/ODBC server or command-line interface where end users or applications can interact with Spark SQL directly to run SQL queries
- It provides integration with Spark Streaming where DStreams can be transformed into structured format and further SQL queries can be executed
- It is capable of caching tables using an in-memory columnar format for faster reads and in-memory data processing
- It supports Schema evolution so that new columns can be added/deleted to the existing schema and Spark SQL still maintains the compatibility between all the versions of the schema

Spark SQL defines the higher level of programming abstraction called **DataFrames**, which is also an extension to the existing RDD API.

DataFrames are the distributed collection of the objects in form the rows and named columns, which is similar to tables in the RDBMS, but with much richer functionality containing all the previously defined features. The DataFrame API is inspired by the concepts of DataFrames in R (http://www.r-tutor.com/r-introduction/data-frame) and Python (http://pandas.pydata.org/pandas-docs/stable/dsintro.html#dataframe).

Let's move ahead and understand how Spark SQL works with the help of an example:

1. As a first step, let's create sample JSON data with basic information about the company's departments such as `Name`, `Employees`, and so on and save this data into a `company.json` file. The JSON file would look similar to this:

    ```
    [
        {
            "Name":"DEPT_A",
            "No_Of_Emp":10,
            "No_Of_Supervisors":2
        },
        {
            "Name":"DEPT_B",
            "No_Of_Emp":12,
            "No_Of_Supervisors":2
        },
        {
            "Name":"DEPT_C",
            "No_Of_Emp":14,
            "No_Of_Supervisors":3
        },
        {
            "Name":"DEPT_D",
            "No_Of_Emp":10,
            "No_Of_Supervisors":1
        },
        {
            "Name":"DEPT_E",
            "No_Of_Emp":20,
            "No_Of_Supervisors":5
        }
    ]
    ```

     You can use any online JSON editor such as `http://codebeautify.org/online-json-editor` to see and edit the data defined in the preceding JSON code.

2. Next, let's extend our Spark-Examples project and create a new package named chapter.six. Within this new package, create a new Scala object and name it ScalaFirstSparkSQL.scala.

3. Next, add the following import statements just below the package declaration:

   ```
   import org.apache.spark.SparkConf
   import org.apache.spark.SparkContext
   import org.apache.spark.sql._
   import org.apache.spark.sql.functions._
   ```

4. Further, in your main method, add the following set of statements to create SQLContext from SparkContext:

   ```
   //Creating Spark Configuration
   val conf = new SparkConf()
   //Setting Application/Job Name
   conf.setAppName("My First Spark SQL")
   //Define Spark Context which we will use to initialize our
   //SQL Context
   val sparkCtx = new SparkContext(conf)
   //Creating SQL Context
   val sqlCtx = new SQLContext(sparkCtx)
   ```

   The SQLContext class or any of its descendants, such as HiveContext for working with Hive tables or CassandraSQLContext for working with Cassandra tables, is the main entry point for accessing all the functionalities of Spark SQL. It allows the creation of DataFrames and also provides functionality to fire SQL queries over DataFrames.

5. Next, we will define the following code to load the JSON file (company.json) using the SQLContext class and further we will also create a data frame:

   ```
   //Define path of your JSON File (company.json) which needs
   //to be processed
   val path = "/home/softwares/spark/data/company.json";
   //Use SQLCOntext and Load the JSON file.
   //This will return the DataFrame which can be further
   //Queried using SQL queries.
   val dataFrame = sqlCtx.jsonFile(path)
   ```

In the preceding piece of code, we are using the jsonFile(...) method for loading the JSON data. There are other utility methods defined by the SQLContext class for reading raw data from filesystem, creating DataFrames from the existing RDD, and many more.

Spark SQL supports two different methods for converting the existing RDDs into DataFrames. The first method uses reflection to infer the schema of an RDD from the given data. This approach leads to more concise code and helps in instances where we already know the schema while writing the Spark application. We have used the same approach in our example.

The second method is through a programmatic interface that allows to construct a schema and then apply it to an existing RDD and finally generate a data frame. This method is more verbose, but provides flexibility and helps in those instances where columns and datatypes are not known until the data is received at runtime.

For a complete list of method exposed by SQLContext, refer to https://spark.apache.org/docs/1.3.0/api/scala/index.html#org.apache.spark.sql.SQLContext.

Once DataFrame is created, we need to register the DataFrame as a temporary table within the SQL context so that we can execute the SQL queries over the registered table. Let's add the following piece of code for registering our DataFrame with our SQL context and name it company:

```
//Register the data as a temporary table within SQL Context
//Temporary table is destroyed as soon as SQL Context is
//destroyed.
dataFrame.registerTempTable("company");
```

We are done! Our JSON data is automatically organized into the table (rows/column) and is ready to accept the SQL queries. Even the datatypes are also inferred from the type of data entered within the JSON file itself.

Now we will start executing the SQL queries on our table, but before this, let's see the schema being created/defined by the SQLContext class:

```
//Printing the Schema of the Data loaded in the Data Frame
dataFrame.printSchema();
```

The execution of the preceding statement will provide results similar to the following screenshot:

```
sumit@localhost $ $SPARK_HOME/bin/spark-submit --class chapter.six.ScalaFirstSparkSQL --master spark://ip-10-69-52-120:7077 Spark-Examples.jar
Spark assembly has been built with Hive, including Datanucleus jars on classpath
15/07/14 01:11:15 WARN NativeCodeLoader: Unable to load native-hadoop library for your platform... using builtin-java classes where applicable
root
 |-- Name: string (nullable = true)
 |-- No_Of_Emp: long (nullable = true)
 |-- No_Of_Supervisors: long (nullable = true)
```

The preceding screenshot shows the schema of the JSON data loaded by Spark SQL. Pretty simple and straightforward, isn't it? Spark SQL has automatically created our schema based on the data defined in our `company.json` file. It has also even defined the datatype of each of the columns.

> We can also define the schema using reflection (https://spark.apache.org/docs/1.3.0/sql-programming-guide.html#inferring-the-schema-using-reflection) or programmatically (https://spark.apache.org/docs/1.3.0/sql-programming-guide.html#inferring-the-schema-using-reflection).

Next, let's execute some SQL queries to see the data stored in the DataFrame, so the first SQL would be to print all the records:

```
//Executing SQL Queries to Print all records in the DataFrame
println("Printing All records")
sqlCtx.sql("Select * from company").collect().foreach(print)
```

The execution of the preceding statement will produce the following results on the console where the driver is executed:

```
sumit@localhost $ $SPARK_HOME/bin/spark-submit --class chapter.six.ScalaFirstSparkSQL --master spark://ip-10-69-52-120:7077 Spark-Examples.jar
Spark assembly has been built with Hive, including Datanucleus jars on classpath
15/07/14 01:11:15 WARN NativeCodeLoader: Unable to load native-hadoop library for your platform... using builtin-java classes where applicable
root
 |-- Name: string (nullable = true)
 |-- No_Of_Emp: long (nullable = true)
 |-- No_Of_Supervisors: long (nullable = true)
Printing All records
[DEPT_A,10,2][DEPT_B,12,2][DEPT_C,14,3][DEPT_D,10,1][DEPT_E,20,5]
```

Next, let's also select only few columns instead of all the records and print the same on the console:

```
//Executing SQL Queries to Print Name and Employees
//in each Department
println("\n Printing Number of Employees in All Departments")
sqlCtx.sql("Select Name, No_Of_Emp from
company").collect().foreach(println)
```

The execution of the preceding statement will produce the following results on the console where the driver is executed:

```
sumit@localhost $ $SPARK_HOME/bin/spark-submit --class chapter.six.ScalaFirstSparkSQL --master spark://ip-10-69-52-120:7077 Spark-Examples.jar
Spark assembly has been built with Hive, including Datanucleus jars on classpath
15/07/14 01:11:15 WARN NativeCodeLoader: Unable to load native-hadoop library for your platform... using builtin-java classes where applicable
root
 |-- Name: string (nullable = true)
 |-- No_Of_Emp: long (nullable = true)
 |-- No_Of_Supervisors: long (nullable = true)
Printing All records
[DEPT_A,10,2][DEPT_B,12,2][DEPT_C,14,3][DEPT_D,10,1][DEPT_E,20,5]
 Printing Number of Employees in All Departments
[DEPT_A,10]
[DEPT_B,12]
[DEPT_C,14]
[DEPT_D,10]
[DEPT_E,20]
```

Integration with Advanced Spark Libraries

Finally, let's do some aggregation and count the total number of employees across the departments:

```
//Using the aggregate function (agg) to print the
//total number of employees in the Company
println("\n Printing Total Number of Employees in Company_X")
val allRec = sqlCtx.sql("Select * from
company").agg(Map("No_Of_Emp"->"sum"))
allRec.collect.foreach ( println )
```

In the preceding piece of code, we are using the agg(...) function and performing the sum of all the employees across the departments, where sum can be replaced by avg, max, min, or count.

The execution of the preceding statement will produce the following results on the console where driver is executed:

```
sumit@localhost $ $SPARK_HOME/bin/spark-submit --class chapter.six.ScalaFirstSparkSQL --master spark://ip-10-69-52-120:7077 Spark-Examples.jar
Spark assembly has been built with Hive, including Datanucleus jars on classpath
15/07/14 01:11:15 WARN NativeCodeLoader: Unable to load native-hadoop library for your platform... using builtin-java classes where applicable
root
 |-- Name: string (nullable = true)
 |-- No_Of_Emp: long (nullable = true)
 |-- No_Of_Supervisors: long (nullable = true)
Printing All records
[DEPT_A,10,2][DEPT_B,12,2][DEPT_C,14,3][DEPT_D,10,1][DEPT_E,20,5]
 Printing Number of Employees in All Departments
[DEPT_A,10]
[DEPT_B,12]
[DEPT_C,14]
[DEPT_D,10]
[DEPT_E,20]

 Printing Total Number of Employees in Company_X
[66]
```

The preceding images shows the results of executing the aggregation on our company.json data.

> For further information on the available functions for performing aggregation, refer to the DataFrame API at https://spark.apache.org/docs/1.3.0/api/scala/index.html#org.apache.spark.sql.DataFrame.

As a last step, we will stop our Spark SQL context by invoking `stop()` function on `SparkContext` – `sparkCtx.stop()`. This is required to let your application notify the master or resource manager to release all the resources allocated to the Spark job. It also ensures the graceful shutdown of the job and avoids any resource leakage that may happen otherwise. Also, as of now there can be only one Spark context active per JVM, and we need to `stop()` the active `SparkContext` before creating a new one.

In this section, we have seen the step-by-step process of using Spark SQL as a standalone program. Though we have considered the JSON files as an example, but we can also leverage Spark SQL with Cassandra (https://github.com/datastax/spark-cassandra-connector/blob/master/doc/2_loading.md), MongoDB (https://github.com/Stratio/spark-mongodb), or Elasticsearch (http://chapeau.freevariable.com/2015/04/elasticsearch-and-spark-1-dot-3.html). Let's move forward toward our next section where we will talk about integrating Spark SQL with Spark Streaming.

Integrating Spark SQL with streams

Let's continue our distributed log processing example and integrate the same with Spark SQL. We will capture the streaming data using Flume and then further perform aggregations using Spark SQL. Refer to the *Data loading from distributed and varied sources* section of *Chapter 3*, *Processing Distributed Log Files in Real Time*, for more details around the use case.

We will enhance our distributed log files processing use case and will leverage Spark SQL for analyzing our Apache log data received/captured with Spark streams in a particular "streaming window". We will first convert our log files into a structured format (DataFrames) and then execute the SQL queries over the structured data for counting the number of distinct type of requests received in a window and finally printing the same on console. Performing data analysis with SQL queries is always preferred because firstly, it provides easy to use functions for aggregations, filtering, merging, and correlation, which is missing in low-level languages such as C or Java; secondly, SQL is easy to learn, adaptable, and widely accepted for performing data analysis as compared to any other programming language.

Here is the overall architecture of our distributed log processing use case after we have integrated Spark Streaming with Spark SQL:

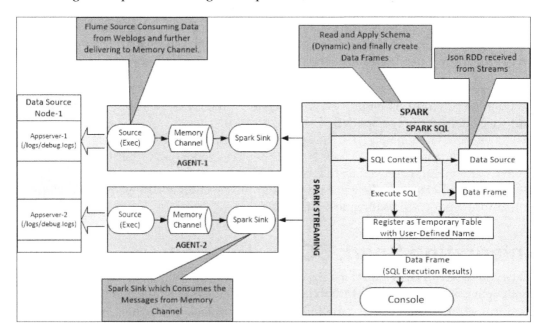

Now that we have implemented Spark Streaming, let's move ahead and perform the following steps for integrating Spark SQL with Spark Streaming:

1. Extend the Spark-Examples project and create a Scala object (ScalaQueryingStreams.scala in package chapter.six).

2. Next, add the following import statements just below the package declaration:
   ```
   import java.net.InetSocketAddress
   import org.apache.spark.SparkConf
   import org.apache.spark.SparkContext
   import org.apache.spark.annotation.Experimental
   import org.apache.spark.sql.SQLContext
   import org.apache.spark.storage.StorageLevel
   import org.apache.spark.streaming.Seconds
   import org.apache.spark.streaming.StreamingContext
   import org.apache.spark.streaming.flume.FlumeUtils
   import chapter.four.ScalaLogAnalyzer
   ```

Chapter 6

3. Download the utility JAR file for converting streams into JSON file from `http://central.maven.org/maven2/com/googlecode/json-simple/json-simple/1.1.1/json-simple-1.1.1.jar`. Save it at `$SPARK_HOME/lib/json-simple-1.1.1.jar` and also add it in your project classpath.

4. Next, edit `chapter.four.ScalaLogAnalyzer` and define a new method, `tansformLogDataIntoJSON(...)`. This new method will parse and convert the streaming data into JSON string:

```
/**
 * Transform the Apache log files and convert them into
 JSON Format
 */
def tansformLogDataIntoJSON(logLine: String): String =
{
  //Pattern which will extract the relevant data from
  //Apache Access Log Files
  val LOG_ENTRY_PATTERN = " """^(\S+) (\S+) (\S+)
  \[([\w:/]+\s[+\-]\d{4})\] "(\S+) (\S+) (\S+)" (\d{3})
  (\S+)"""
  val PATTERN = Pattern.compile(LOG_ENTRY_PATTERN)
  val matcher = PATTERN.matcher(logLine)

  //Matching the pattern for the each line of the Apache
  //access Log file
  if (!matcher.find()) {
    System.out.println("Cannot parse logline" + logLine)
  }

  //Creating the JSON Formatted String from the Map
  import scala.collection.JavaConversions._
  val obj = new
  JSONObject(mapAsJavaMap(createDataMap(matcher)))
  val json = obj.toJSONString()

  println("JSON DATA new One - ", json )
  return json
}
```

Integration with Advanced Spark Libraries

5. Next, edit `ScalaQueryingStreams` and add the following piece of code:

```
def main(args: Array[String]) {

  //Creating Spark Configuration
  val conf = new SparkConf()
  conf.setAppName("Integrating Spark SQL")
  //Define Spark Context which we will use to initialize
  //our SQL Context
  val sparkCtx = new SparkContext(conf)
  //Retrieving Streaming Context from Spark Context
  val streamCtx = new StreamingContext(sparkCtx,
  Seconds(10))

  //Defining Host for receiving data from Flume Sink
  var addresses = new Array[InetSocketAddress](1);
  addresses(0) = new InetSocketAddress("localhost", 4949)
  //Creating Flume Polling Stream
  val flumeStream = FlumeUtils.createPollingStream
  (streamCtx, addresses,
  StorageLevel.MEMORY_AND_DISK_SER_2, 1000, 1)

  //Utility class for Transforming Log Data
  val transformLog = new ScalaLogAnalyzer()

  //Invoking map() operation to convert the log data into
  //RDD of JSON Formatted String
  val newDstream = flumeStream.map { x =>
  transformLog.tansformLogDataIntoJSON(new
  String(x.event.getBody().array())) }

  //Defining Window Operation, So that we can execute SQL
  //Query on data received within a particular Window
  val wStream = newDstream.window(Seconds(40),
  Seconds(20))

  //Creating SQL DataFrame for each of the RDD's
  wStream.foreachRDD { rdd =>
    //Getting the SQL Context from Utility Method which
    //provides Singleton Instance of SQL Context
    val sqlCtx = getInstance(sparkCtx)
    //Converting JSON RDD into the SQL DataFrame by using
    //jsonRDD() function
```

```
      val df = sqlCtx.jsonRDD(rdd)
      //creating and Registering the Temporary table for
      //the Converting DataFrame into table for further
      //Querying
      df.registerTempTable("apacheLogData")

      //Print the Schema
      println("Here is the Schema of your Log
      Files............")
      df.printSchema()
      //Executing the Query to get the total count of
      //different HTTP Response Code in the Data Frame
      val logDataFrame = sqlCtx.sql("select method,
      count(*) as total from apacheLogData
      group by method")
      //Finally printing the results on the Console
      println("Total Number of Requests............. ")
      logDataFrame.show()
    }

    streamCtx.start();
    streamCtx.awaitTermination();
  }

  //Defining Singleton SQLContext variable
  @transient private var instance: SQLContext = null

  //Lazy initialization of SQL Context
  def getInstance(sparkContext: SparkContext): SQLContext =
  synchronized {
    if (instance == null) {
      instance = new SQLContext(sparkContext)
    }
    instance
  }
```

We are done! Our Spark streams are converted into Spark SQL DataFrames and we have also written the SQL queries for querying the data.

Now let's move forward and perform the following steps for executing the preceding piece of code:

1. Edit `$SPARK_HOME/conf/spark-defaults.conf` and append the value of the `spark.driver.extraClassPath` and `spark.executor.extraClassPath` parameters with the location of your JSON jar file, that is, `$SPARK_HOME/lib/json-simple-1.1.1.jar`.

2. Assuming your Spark cluster is down, let's bring up our Spark cluster by performing the following steps:

   1. Bring up our Spark master by executing the following command on your Linux Console:

      `$SPARK_HOME/sbin/start-master.sh`

   2. Next, bring up our Spark worker by executing the following command on your Linux console:

      `$SPARK_HOME/bin/spark-class org.apache.spark.deploy.worker.Worker <Spark Master URL> &`

   3. Execute the following command to bring up our Flume agent:

      `$FLUME_HOME/bin/flume-ng agent --conf conf --conf-file $FLUME_HOME/conf/spark-flume.conf --name a1 &`

   4. Compile and export our `Spark-Examples` project and create a JAR file named `Spark-Examples.jar`.

   5. Execute the following command to simulate the log generation in real time from the folder where our `Spark-Examples.jar` file is saved:

      `java -classpath Spark-Examples.jar chapter.four.SampleLogGenerator <path of file for saving log file> <path of "access_log" file> &`

   Our Spark environment is ready. Next, we will move forward and execute our Spark job for consuming events and transform them into DataFrames.

3. Execute the following command on Linux console for executing our Spark job:

   `$SPARK_HOME/bin/spark-submit --class chapter.six.ScalaQueryingStreams --master <Spark-master-URL> Spark-Examples.jar`

   As soon as we execute the preceding command, the logs would be consumed and we would get the results on the console that would look similar to the following screenshot:

```
sumit@localhost $ $SPARK_HOME/bin/spark-submit --class chapter.six.ScalaQueryingStreams --master spark://ip-10-69-52-120:7077 Spark-Examples.jar
Spark assembly has been built with Hive, including Datanucleus jars on classpath
15/07/14 03:23:56 WARN NativeCodeLoader: Unable to load native-hadoop library for your platform... using builtin-java classes where applicable
15/07/14 03:24:01 INFO ipc.NettyServer: [id: 0xb3255f66, /127.0.0.1:44870 => /127.0.0.1:4949] OPEN
15/07/14 03:24:01 INFO ipc.NettyServer: [id: 0xb3255f66, /127.0.0.1:44870 => /127.0.0.1:4949] BOUND: /127.0.0.1:4949
15/07/14 03:24:01 INFO ipc.NettyServer: [id: 0xb3255f66, /127.0.0.1:44870 => /127.0.0.1:4949] CONNECTED: /127.0.0.1:44870
15/07/14 03:24:07 WARN sink.TransactionProcessor: Spark could not commit transaction, NACK received. Rolling back transaction.
15/07/14 03:24:07 WARN sink.TransactionProcessor: Spark was unable to successfully process the events. Transaction is being rolled back.
Here is the Schema of your Log Files...........
root
 |-- IP: string (nullable = true)
 |-- client: string (nullable = true)
 |-- date: string (nullable = true)
 |-- method: string (nullable = true)
 |-- protocol: string (nullable = true)
 |-- request: string (nullable = true)
 |-- respCode: string (nullable = true)
 |-- size: string (nullable = true)
 |-- user: string (nullable = true)

Total Number of Requests............
method total
GET     6
Here is the Schema of your Log Files...........
root
 |-- IP: string (nullable = true)
 |-- client: string (nullable = true)
 |-- date: string (nullable = true)
 |-- method: string (nullable = true)
 |-- protocol: string (nullable = true)
 |-- request: string (nullable = true)
 |-- respCode: string (nullable = true)
 |-- size: string (nullable = true)
 |-- user: string (nullable = true)

Total Number of Requests............
method total
GET     12
Here is the Schema of your Log Files...........
root
 |-- IP: string (nullable = true)
 |-- client: string (nullable = true)
 |-- date: string (nullable = true)
 |-- method: string (nullable = true)
 |-- protocol: string (nullable = true)
 |-- request: string (nullable = true)
 |-- respCode: string (nullable = true)
 |-- size: string (nullable = true)
 |-- user: string (nullable = true)
```

The preceding screenshot shows the Schema of our log data captured in each stream of events and at the same time, it also prints the total number of unique request being served by the web server from which the log files are analyzed.

In this section, we have discussed about the integration of Spark streams with Spark SQL. We captured the streaming log data and converted them into Spark SQL DataFrames and then further executed the queries for getting the distinct requests.

Let's move forward and see the integration of Spark Streaming with Spark GraphX.

Graph analysis – Spark GraphX

In today's world, nothing exists in isolation; everything is connected with each other and graphs are one of most efficient and widely used methodologies for representing data in the form of nodes and the relationships between them.

Graphs and graph analysis have been used by data scientists and researchers to uncover new, interesting, and hidden patterns in datasets which would otherwise be difficult to see with the naked eye.

The social networks, network management, genealogy, public transport links, and road maps are all examples of such complex use cases that require a generic data structure to elegantly and efficiently represent the data in form of relationships, and at the same time, making them easy to query in a highly-accessible manner.

Though there is no universal accepted definition of graphs, but as per Wikipedia (http://en.wikipedia.org/wiki/Graph_(abstract_data_type)):

> *"A graph data structure consists of a finite (and possibly mutable) set of nodes or vertices, together with a set of ordered pairs of these nodes (or, in some cases, a set of unordered pairs). These pairs are known as edges or arcs. As in mathematics, an edge (x, y) is said to point or go from x to y. The nodes may be part of the graph structure, or may be external entities represented by integer indices or references."*

Graph analysis or graph data structure is not a new terminology, and there have been open source systems such as Apache Giraph (http://giraph.apache.org/), GraphLab (https://en.wikipedia.org/wiki/GraphLab), and Google's Pregel (http://kowshik.github.io/JPregel/pregel_paper.pdf), which exposes the specialized APIs for simplifying graph programming. However, all these systems do not provide efficient, fault tolerant and parallel or distributed computations/transformations of graphs.

Spark introduced a unified approach for storing, processing, and transforming the data in form of graphs over its distributed in-memory data processing platform, which is known as GraphX.

GraphX is developed on the concepts of **property graph model** where data is represented in form of vertices and edges:

- **Vertices**: These are the also known as **nodes** that represent the entity and contain the attributes of that entity
- **Edges**: These are used to define the relationships or connection between vertices/nodes in the form of lines, defining the direction of relationships

Let's take an example of school where we have subjects, teachers and students and consider the following relationships between the three:

- A school provides specialization in three subjects—Science, Math, and English
- There are three specialized teachers for teaching these subjects:
    - Mary teaches Science
    - Leena teaches English
    - Kate teaches Math

- There are two students enrolled into each subject:
    - Science: Joseph and Adam
    - Math: Ram and Jessica
    - English: Brooks and Camily

Representing it graphically in the form of graphs would look something similar to the following illustration:

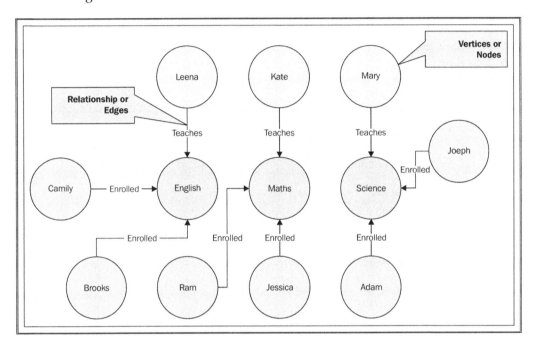

The preceding image illustrates the visual representation of subject versus student versus teacher relationship.

Introduction to the GraphX API

In this section, we will discuss about the usage of the various APIs exposed by Spark GraphX.

Let's move forward and extend the example of students/teacher/subjects and define the same model using Spark GraphX for understanding the various APIs exposed by Spark GraphX.

Integration with Advanced Spark Libraries

Perform the following steps for defining the Graph model using the Spark GraphX API:

1. Let's extend our `Spark-Examples` project and create a new package, `chapter.six`, and under this package, create a new Scala object and name it `ScalaSparkGraphx.scala`.

2. Edit `ScalaSparkGraphx.scala` and add the following packages just below the package declaration:

   ```
   import org.apache.spark.SparkConf
   import org.apache.spark.SparkContext
   import org.apache.spark.graphx._
   import org.apache.spark.graphx.Graph.graphToGraphOps
   import org.apache.spark.rdd.RDD
   ```

3. Next, add a `main` method and add the following piece of code within it:

   ```
   def main(args: Array[String]) {

     //Creating Spark Configuration
     val conf = new SparkConf()
     conf.setAppName("My First Spark GraphX ")
     //Define Spark Context
     val sparkCtx = new SparkContext(conf)

     //Define Vertices/Nodes for Subjects
     //"parallelize()" is used to distribute a local Scala
     //collection to form an RDD.
     //It acts lazily, i.e.
     //if a mutable collection is altered after invoking
     //"parallelize" but before
     //any invocation to the RDD operation then your resultant
     //RDD will contain modified collection.
     val subjects: RDD[(VertexId, (String))] =
     sparkCtx.parallelize(Array((1L, ("English")), (2L,
     ("Math")),(3L, ("Science"))))

     //Define Vertices/Nodes for Teachers
     val teachers: RDD[(VertexId, (String))] =
     sparkCtx.parallelize(Array((4L, ("Leena")), (5L,
     ("Kate")),(6L, ("Mary"))))
     //Define Vertices/Nodes for Students
     val students: RDD[(VertexId, (String))] =
     sparkCtx.parallelize(Array((7L, ("Adam")), (8L,
     ("Joseph")),(9L, ("Jessica")),(10L, ("Ram")),(11L,
     ("brooks")),(12L, ("Camily"))))
   ```

```
//Join all Vertices and create 1 Vertex
val vertices = subjects.union(teachers).union(students)

//Define Edges/Relationships between Subject versus
//Teachers
val subjectsVSteachers: RDD[Edge[String]] =
sparkCtx.parallelize(Array(Edge(4L,1L, "teaches"),
Edge(5L,2L, "teaches"),Edge(6L, 3L, "teaches")))

//Define Edges/Relationships between Subject vs Students
val subjectsVSstudents: RDD[Edge[String]] =
sparkCtx.parallelize(
Array(Edge(7L, 3L, "Enrolled"), Edge(8L, 3L,
"Enrolled"),Edge(9L, 2L, "Enrolled"),Edge(10L, 2L,
"Enrolled"),Edge(11L, 1L, "Enrolled"),
Edge(12L, 1L, "Enrolled")))
//Join all Edges and create 1 Edge
val edges = subjectsVSteachers.union(subjectsVSstudents)
//Define Object of Graph
val graph = Graph(vertices, edges)

//Print total number of Vertices and Edges
println("Total vertices = " + graph.vertices.count()+",
Total Edges = "+graph.edges.count())

//Print Students and Teachers associated with Subject =
//Math
println("Students and Teachers associated with Math....")
//EdgeTriplet represents an edge along with the vertex
//attributes of its neighboring vertices.triplets
graph.triplets.filter(f=>f.dstAttr.equalsIgnoreCase
("Math")).collect().foreach(println)

sparkCtx.stop()
}
```

We are done! Let's move ahead and execute the preceding piece of code and see the results on Spark console.

4. Compile your Scala code, export it as JAR file, and name it Spark-Examples.jar.
5. Assuming your Spark cluster is up and running, open your Linux console and browse $SPARK_HOME and execute the following command for deploying your Spark Streaming job:

```
$SPARK_HOME/bin/spark-submit --class chapter.six.ScalaSparkGraphx
--master <SPARK-MASTER-URL> Spark-Examples.jar
```

Integration with Advanced Spark Libraries

As soon as you click on *Enter* and execute the preceding command, you will see that your graph is created; the queries will be executed on your graph data model and will produce the results, which would look something similar to the following screenshot showing the output of the queries:

```
sumit@localhost $ $SPARK_HOME/bin/spark-submit --class chapter.six.ScalaSparkGraphx --master spark://ip-10-69-52-120:7077 Spark-Examples.jar
Spark assembly has been built with Hive, including Datanucleus jars on classpath
15/07/17 10:50:55 WARN NativeCodeLoader: Unable to load native-hadoop library for your platform... using builtin-java classes where applicable
Total vertices = 12, Total Edges = 9
Students and Teachers associated with Math....
((5,Kate),(2,Math),teaches)
((9,Jessica),(2,Math),Enrolled)
((10,Ram),(2,Math),Enrolled)
```

For more information on Spark GraphX, refer to https://spark.apache.org/docs/1.3.0/graphx-programming-guide.html.

Integration with Spark Streaming

Spark GraphX does not provide any direct integration with Spark Streaming, but GraphX, at high level, extends the Spark RDD by introducing the resilient distributed property graph: a directed multigraph with properties attached to each vertex and edge.

Same goes for DStreams too, where they are nothing more than a continuous sequence of RDDs (of the same type), representing a continuous stream of data. So, the integration between streaming and GraphX is possible via RDDs. In DStream, RDDs can be in transformed into vertices and edges and further, they can be directly used to create the Graphs for every chunk of data received or for the data received within a window.

For example, consider our distributed log processing/analysis example, which was introduced in *Chapter 3, Processing Distributed Log Files in Real Time*, where we consume the events from distributed sources using Spark-Flume sink. We can further extend this example where we can create the graph of each chunk of data received from the Flume streams in a window where each line of event/log can be unwrapped/parsed and all the attributes can be converted into vertices and further, all the vertices can be connected to one of the attributes (let's say, "IP"), resulting in the edges.

This is exactly what we will be doing where we will create the graph of all the requests received in a particular time window (as defined by our streams) from distinct IP addresses. This will not only help us to analyze the various requests received from different users/IPs, but at the same time, we can also analyze the popular pages by the number of time they have been requested by the various users in a specific time window.

The following illustration shows the high-level architecture of our distributed log processing example, which is extended and converted into Graphs using the Spark GraphX APIs:

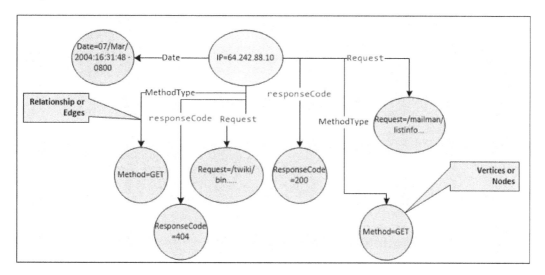

Graphs are really helpful for analyzing the hidden patterns and trends that are not possible to uncover with the naked eye, not even with SQL. Even in our distributed log processing example, we can uncover the hidden patterns or changing trends by evaluating the URLs requested by the various users over the period of time. This would tell us the changing behaviors of our prospective users or also the increase or decline in activities of the various users visiting our website.

Let's move forward and perform the following steps for extending our distributed log processing example and convert chunk of log events into graphs:

1. Extend our `Spark-Examples` project and create a Scala object: `ScalaCreateStreamingGraphs.scala` in package `chapter.six`.

2. Next, edit `ScalaCreateStreamingGraphs.scala` and add the following `import` statements:

```
import org.apache.spark._
import org.apache.spark.streaming._
import org.apache.spark.graphx._
import org.apache.spark.rdd.RDD
import org.apache.spark.streaming.flume._
import org.apache.spark.storage.StorageLevel
import org.apache.spark.rdd._
import org.apache.spark.streaming.dstream._
import java.net.InetSocketAddress
import chapter.four.ScalaLogAnalyzer
```

3. Next, open and edit `chapter.four.ScalaLogAnalyzer` and define the utility method, `transformIntoGraph(...)`, which process the array of Flume events and creates vertices and edges:

```
/**
* Utility method for transforming Flume Events into
Sequence of Vertices and Edges
*/
def transformIntoGraph(eventArr: Array[SparkFlumeEvent]):
Tuple2[Set[(VertexId, (String))],Set[Edge[String]]] = {
  println("Start Transformation........")
  //Defining mutable Sets for holding the Vertices and
  //Edges
  val verticesSet: scala.collection.mutable.
  Set[(VertexId,String)] = scala.collection.mutable.Set()
  val edgesSet: scala.collection.mutable.
  Set[Edge[String]] = scala.collection.mutable.Set()

  //Creating Map of IP and Vertices ID,
  //so that we create Edges to the same IP
  var ipMap:Map[String,Long] = Map()

  //Looping over the Array of Flume Events
  for(event<-eventArr){
    //Get the Line of Log and Transform into
    //Attribute Map
    val eventAttrMap = tansfromLogData(new
    String(event.event.getBody().array()))

    //Using Random function for defining Unique Vertices
    //ID's
    //Creating Vertices for IP
    //Creating new or Getting existing VertexID for
    //IP coming from Events
    val ip_verticeID:Long =
    if(ipMap.contains(eventAttrMap.get("IP").get)){
      ipMap.get(eventAttrMap.get("IP").get).get

    }
    else{
      //Using Random function for defining Unique
      //Vertex ID's
      val id = Random.nextLong()
      //Add to the Map
      ipMap+= (eventAttrMap.get("IP").get -> id)
      //Return the Value
      id
    }
    //Add Vertex for IP
```

```
            verticesSet+=((ip_verticeID,
            "IP="+eventAttrMap.get("IP")))
            //Creating Vertex for Request
            val request_verticeID = Random.nextLong()
            verticesSet+=((request_verticeID,
            "Request="+eventAttrMap.get("request")))
            //Creating Vertice for Date
            val date_verticeID = Random.nextLong()
            verticesSet+=((date_verticeID,
            "Date="+eventAttrMap.get("date")))
            //Creating Vertice for Method
            val method_verticeID = Random.nextLong()
            verticesSet+=((method_verticeID,
            "Method="+eventAttrMap.get("method")))
            //Creating Vertice for Response Code
            val respCode_verticeID = Random.nextLong()
            verticesSet+=((respCode_verticeID,
            "ResponseCode="+eventAttrMap.get("respCode")))

            //Defining Edges. All parameters are
            //in relation to the User IP
            edgesSet.+=(Edge(ip_verticeID,
            request_verticeID,"Request")).+=
            (Edge(ip_verticeID,date_verticeID,"date"))
            edgesSet.+=(Edge(ip_verticeID,method_verticeID,
            "methodType")).+=(Edge(ip_verticeID,
            respCode_verticeID,"responseCode")
          }
          println("End Transformation........")

          //Finally Return the Tuple of 2 Set containing Vertices
    and Edges
          return   (verticesSet,edgesSet)

      }
```

4. Edit `ScalaCreateStreamingGraphs.scala` and within the `main` method, add the following piece of code:

```
def main(args: Array[String]) {

  //Creating Spark Configuration
  val conf = new SparkConf()
  conf.setAppName("Integrating Spark Streaming with
  GraphX")
  //Define Spark Context which we will use to initialize
  //our SQL Context
  val sparkCtx = new SparkContext(conf)
  //Retrieving Streaming Context from Spark Context
```

Integration with Advanced Spark Libraries

```scala
val streamCtx = new StreamingContext(sparkCtx,
Seconds(10))
//Address of Flume Sink
var addresses = new Array[InetSocketAddress](1);
addresses(0) = new InetSocketAddress("localhost",4949)
//Creating a Stream
val flumeStream = FlumeUtils.createPollingStream
(streamCtx,addresses,StorageLevel.
MEMORY_AND_DISK_SER_2,1000,1)
//Define Window Function to collect Event for a certain
//Duration
val newDstream =
flumeStream.window(Seconds(40),Seconds(20))

//Define Utility class for Transforming Log Data
val transformLog = new ScalaLogAnalyzer()
//Create Graphs for each RDD
val graphStream = newDstream.foreachRDD { x =>
  //Invoke utility Method for Transforming Events into
  //Graphs Vertices and Edges
  //Wrapped in a Mutable Seq
  val tuple = transformLog.
  transformIntoGraph(x.collect())
  println("Creating Graphs Now.................")
  //Define Vertices
  val vertices:RDD[(VertexId, (String))] =
  sparkCtx.parallelize(tuple._1.toSeq)
  //Define Edges
  val edges:RDD[Edge[String]] = sparkCtx.
  parallelize(tuple._2.toSeq)
  //Create or Initialize Graph
  val graph = Graph(vertices,edges)
  //Print total number of Vertices and Edges in
  the Graph
  println("Total vertices = " +
  graph.vertices.count()+", Total Edges =
  "+graph.edges.count())
  //Printing All Vertices in the Graph
  graph.vertices.collect().iterator.
  foreach(f=>println("Vertex-ID = "+f._1+",
  Vertex-Name = "+f._2))
  //Printing Requests from Distinct IP's in this Window
  println("Printing Requests from Distinct IP's in
  this Window")
  println("Here is the Count = "+graph.
  vertices.filter ( x =>
  x._2.startsWith("IP")).count())
  println("Here is the Distinct IP's = ")
```

```
            graph.vertices.filter ( x => x._2.startsWith("IP")).
            collect.foreach(ip=>println(ip._2))

            //Printing count of Distinct URL requested
            in this Window
            println("Printing count of Distinct URL
            requested in this Window")
            val filteredRDD = graph.vertices.filter
            ( x => x._2.startsWith("Request=")).map(x =>
            (x._2, 1)).reduceByKey(_+_)
            filteredRDD.collect.foreach(url=>println
            (url._1+" = " + url._2))
        }

        streamCtx.start();
        streamCtx.awaitTermination();
    }
```

Our integration is completed. Now, for every chunk of data received, the preceding piece of code will create a graph that can be further analyzed or queried. Now, for running the example, first we have to bring up our cluster and follow the same steps as we did in the previous section, *Integrating Spark SQL with streams*. Once the cluster is up and running we can execute the following command on our Linux console:

$SPARK_HOME/bin/spark-submit --class chapter.six. ScalaCreateStreamingGraphs --master <Spark-master-URL> Spark-Examples.jar

As soon as we execute this command, the logs would be consumed and we would get the results on the console that would look similar to the following screenshot:

```
sumit@localhost $ $SPARK_HOME/bin/spark-submit --class chapter.six.ScalaCreateStreamingGraphs --master spark://ip-10-157-113-61:7077 Spark-Examples.jar
Spark assembly has been built with Hive, including Datanucleus jars on classpath
15/08/26 02:55:10 WARN NativeCodeLoader: unable to load native-hadoop library for your platform... using builtin-java classes where applicable
15/08/26 02:55:16 INFO ipc.NettyServer: [id: 0x3b0fccf0, /127.0.0.1:37014 => /127.0.0.1:4949] OPEN
15/08/26 02:55:16 INFO ipc.NettyServer: [id: 0x3b0fccf0, /127.0.0.1:37014 => /127.0.0.1:4949] BOUND: /127.0.0.1:4949
15/08/26 02:55:16 INFO ipc.NettyServer: [id: 0x3b0fccf0, /127.0.0.1:37014 => /127.0.0.1:4949] CONNECTED: /127.0.0.1:37014
Start Transformation.....
End Transformation........
Creating Graphs Now.............
Total vertices = 25, Total edges = 24
Vertex-ID = -1532384247150155171, Vertex-Name = Date=Some(07/Mar/2004:16:06:51 -0800)
Vertex-ID = -5418176438795731804, Vertex-Name = ResponseCode=Some(200)
Vertex-ID = 6109152814336663741, Vertex-Name = Method=Some(GET)
Vertex-ID = -8167173914399604184, Vertex-Name = Date=Some(07/Mar/2004:16:24:16 -0800)
Vertex-ID = -1461351152372048920, Vertex-Name = Date=Some(07/Mar/2004:16:10:02 -0800)
Vertex-ID = 741344983461260864, Vertex-Name = Request=Some(/twiki/bin/view/Main/PeterThoeny)
Vertex-ID = 3672188992249921222, Vertex-Name = ResponseCode=Some(200)
Vertex-ID = -8380893481359571383, Vertex-Name = Method=Some(GET)
Vertex-ID = 2232910925110990062, Vertex-Name = Method=Some(GET)
Vertex-ID = 5889234281880079157, Vertex-Name = ResponseCode=Some(200)
Vertex-ID = 7493449142301548152, Vertex-Name = Request=Some(/mailman/listinfo/hsdivision)
Vertex-ID = 6465687430844502981, Vertex-Name = Date=Some(07/Mar/2004:16:05:49 -0800)
Vertex-ID = -2107191155499873541, Vertex-Name = Method=Some(GET)
Vertex-ID = 4580235862311159076, Vertex-Name = Date=Some(07/Mar/2004:16:24:16 -0800)
Vertex-ID = -7894905428967736821, Vertex-Name = Request=Some(/twiki/bin/rdiff/Twiki/NewUserTemplate?rev1=1.3&rev2=1.2)
Vertex-ID = -8300291579112802156, Vertex-Name = Date=Some(07/Mar/2004:16:24:16 -0800)
Vertex-ID = 3672188992249921222, Vertex-Name = ResponseCode=Some(200)
Vertex-ID = -1928305135849453897, Vertex-Name = Method=Some(GET)
Vertex-ID = 8762283324744869047, Vertex-Name = ResponseCode=Some(200)
Vertex-ID = 5918795085635250219, Vertex-Name = ResponseCode=Some(200)
Vertex-ID = 2188202781662311568, Vertex-Name = Method=Some(GET)
Vertex-ID = -3472534429404585923, Vertex-Name = ResponseCode=Some(401)
Vertex-ID = -8209573003173141377, Vertex-Name = IP=Some(64.242.88.10)
Vertex-ID = -7370385737099862881, Vertex-Name = Request=Some(/twiki/bin/edit/Main/Double_bounce_sender?topicparent=Main.ConfigurationVariables)
Vertex-ID = 8184300095548333942, Vertex-Name = Request=Some(/twiki/bin/view/Main/PeterThoeny)
Vertex-ID = 9063684431438294183, Vertex-Name = Request=Some(/twiki/bin/view/Main/PeterThoeny)
Printing Requests from Distinct IP's in this window
Here is the Count = 1
Here is the Distinct IP's =
IP=Some(64.242.88.10)
Printing Distinct URL requested in this window
Request=Some(/twiki/bin/view/Main/PeterThoeny) = 3
Request=Some(/mailman/listinfo/hsdivision) = 1
Request=Some(/twiki/bin/edit/Main/Double_bounce_sender?topicparent=Main.ConfigurationVariables) = 1
Request=Some(/twiki/bin/rdiff/Twiki/NewUserTemplate?rev1=1.3&rev2=1.2) = 1
```

Integration with Advanced Spark Libraries

The preceding screenshot shows the number of vertices and edges created after parsing the log data and at the same time, it prints `Vertex-ID` and `Vertex-Name` on the console. It also shows the request received from distinct IP addresses and the count of distinct URLs.

> We can extend our example, where we can persist the vertices and edges in Spark memory or on some persistent storage area such as HDFS across the streaming time windows and further join them with the data received in the subsequent time windows to create one single graph for each user/IP address.

In this section, we discussed about the integration of Spark streams with Spark GraphX. We captured the streaming log data and converted them into graphs using the Spark GraphX APIs and then further executed the few basic queries.

Summary

In this chapter, we discussed about the integration of advanced spark libraries like Spark SQL and Spark GraphX with Spark Streaming. We also extended our distributed log analysis example and applied the Spark SQL and GraphX APIs over the distributed events received in near real time.

In the next chapter, we will discuss the various deployment aspects of Spark and Spark Streaming.

7
Deploying in Production

If you can't get software into the hands of your users, then what is its value? Zero, isn't it?

That's exactly the reason that every type of software at the end of the day needs to be deployed in production, and that's where the real challenge is!

Production deployment is a complex endeavor, which is one of the most critical aspects of any **software development life cycle** (**SDLC**). Development teams are constantly striving to deploy software in such a way that various enterprise concerns such as maintenance, backups, restores, and disaster recovery are easier to perform and flexible enough to scale and accommodate the future needs. Architects/developers envision their production environment/deployments from day-1 (sometimes, even before) of the project kick-off.

Though application software or frameworks do provide various deployment options, which one to use and how to use will largely depend upon the usage of software by the end users.

For example, you may have strict SLAs for consuming/processing near real-time feeds in comparison to your batch processes. So, your deployments will be configured and optimized in such a manner that consumption or processing of your near real-time feeds will always take precedence over your batch processes.

We should also remember that the production deployments are evolving and would change over the period of time due to many reasons, such as new versions, innovation in hardware, change in user behavior, business dynamics, and others.

Deploying in Production

The same is true for Spark and Spark-based applications, but lately, there have been a lot of rumors about Spark and its production readiness and if this was not enough, we must have heard people categorizing Spark and Mesos as one and the same. Funny, isn't it?

That's true that the idea of Spark was originated while working on Apache Mesos at the University of California, Berkeley, but both are not the same. Apache Mesos is an open source cluster manager that provides efficient resource isolation and sharing across the distributed applications or frameworks, while Apache Spark is an open source cluster computing framework that requires an efficient cluster manager such as Apache Mesos or YARN and a distributed storage System such as HDFS. In a nutshell, Apache Mesos and Spark are not the same, they are different!

As far as production readiness is concerned, there are more than 80 companies already using Spark in production and the list is growing day by day. You can find more about it at `https://cwiki.apache.org/confluence/display/SPARK/Powered+By+Spark`.

Let's move forward and read more about the deployment aspects and the concerns of Spark and Spark-based applications.

This chapter will cover the following points:

- Spark deployment models
- High availability and fault tolerance
- Monitoring streaming jobs

Spark deployment models

Spark is developed as a framework that can be further deployed on the various distributed cluster computing frameworks such as Hadoop/YARN, Apache Mesos, and standalone too.

Spark specifies the integration hooks that can be extended and integrated in such a manner that our Spark applications can leverage the cluster manager of other distributed cluster computing frameworks that may provide efficient resource isolation and sharing across distributed applications.

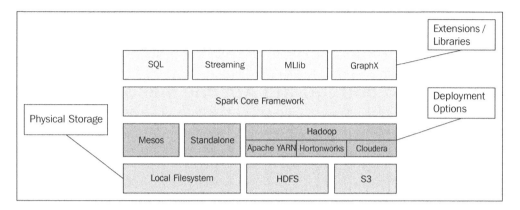

The preceding illustration shows different deployment options available for deploying Spark and Spark-based applications.

In the previous chapters, we executed our Spark and Spark Streaming jobs in standalone cluster mode provided by Spark itself.

Let's move forward and discuss the deployment steps for deploying our Spark applications on other cluster computing frameworks such as Apache Mesos and YARN.

Deploying on Apache Mesos

Apache Mesos (http://mesos.apache.org/) is a cluster manager that provides efficient resource isolation and sharing across distributed applications or frameworks. It can run Hadoop, MPI, Hypertable, Spark, and other frameworks on a dynamically shared pool of nodes.

Apache Mesos and Spark are closely related to each other (but they are not same). The story started back in 2009 when Mesos was ready and there were thoughts going on about the ideas/framework that can be developed on top of Mesos, and that's exactly how Spark was born.

The objective was to showcase how easy it was to build a framework from scratch in Mesos and at the same time, the target was to support interactive and iterative computations like machine learning and also provide ad hoc querying.

Mesos completely abstracts out compute resources such as CPU, memory, and others from the machines (physical or virtual) and enables the fault tolerant and elastic distribution of the compute resources. Mesos maintains a pool of compute resources that are allocated to the applications as per the demand/request.

Installing and configuring Apache Mesos

Let's move ahead and install Apache Mesos and its dependent components. Then, we will also deploy and execute our Spark examples on the same cluster.

Perform the following steps for installing Apache Mesos and its dependent components on Linux:

1. Execute the following commands to install Apache Maven (https://maven.apache.org/):

   ```
   wget http://mirror.olnevhost.net/pub/apache/maven/maven-3/3.0.5/binaries/apache-maven-3.0.5-bin.tar.gz
   ```
   ```
   tar xvf apache-maven-3.0.5-bin.tar.gz
   ```
   ```
   export M2_HOME=<path of installation Directory>
   ```
   ```
   export M2=$M2_HOME/bin
   ```
   ```
   export PATH=$M2:$PATH
   ```

2. Next, execute the following Linux commands for installing Apache Mesos dependencies:

   ```
   sudo yum install  -y autoconf libtool
   ```
   ```
   sudo yum install -y build-essential python-dev python-boto libcurl4-nss-dev libsasl2-dev maven libapr1-dev libsvn-dev
   ```
   ```
   sudo yum groupinstall -y "Development Tools"
   ```
   ```
   sudo yum install -y python-devel  zlib-devel libcurl-devel openssl-devel cyrus-sasl-devel cyrus-sasl-md5 apr-devel subversion-devel apr-util-devel
   ```

3. Now, execute the following commands on your Linux console for downloading and extracting Apache Mesos; let's refer the directory, in which we will extract Mesos, as <MESOS_HOME>:

   ```
   wget http://www.apache.org/dist/mesos/0.22.1/mesos-0.22.1.tar.gz
   tar -zxf mesos-0.22.1.tar.gz
   ```

4. Then, execute the following Linux commands for compiling the Mesos libraries and creating a build:

   ```
   mkdir <MESOS_HOME>/build
   cd <MESOS_HOME>/build
   ../configure
   Make
   ```

   > Replace <MESOS_HOME> with the actual path of the directory in all the instances.

   We are done with the installation. Our Apache Mesos binaries are compiled and are ready to be used.

5. In <MESOS_HOME>/build/, execute the following command on Linux console to start the Mesos master:

   ```
   mkdir workdir
   ./bin/mesos-master.sh --cluster='Spark Cluster' --work_dir=<MESOS_HOME>/build/workdir &
   ```

6. Next, bring up the Mesos slave by executing the following command:

   ```
   /bin/mesos-slave.sh --master=<IP of the server>:5050
   ```

   The <IP of the server> is the IP of your local machine that is binded by Mesos master. In case you have more than one network interfaces, you can also explicitly specify the IP by providing the --ip=<IP-Address> option while bringing up your master. Alternatively, you can also see this IP on the master UI (http://<host-name>:5050).

Deploying in Production

As soon as you execute the preceding command, your Apache Mesos cluster is up and running with one master and one slave and is ready to accept the request. We can browse the Mesos master at http://<host-name>:5050. It would look similar to the following screenshot:

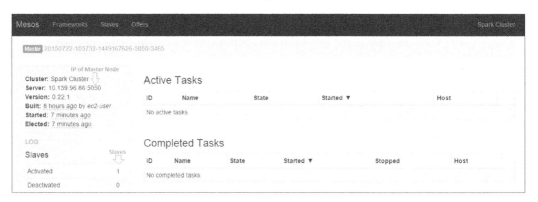

The preceding screenshot shows the Mesos master UI. It is also known as admin console or monitoring console where we can monitor all the active and completed jobs/tasks and of course the health of the cluster can also be monitored from the same UI.

 For more information on Mesos, refer to the official documentation at http://mesos.apache.org/documentation/latest/.

Integrating and executing Spark applications on Apache Mesos

Perform the following steps for integrating and submitting Spark applications on Apache Mesos cluster:

1. Execute the following command on your Linux console for setting the environment variable for Mesos native library, libmesos.so, which will be used by Java runtime for executing our Spark jobs:

   `export MESOS_NATIVE_JAVA_LIBRARY=<Path >`

   By default, libmesos.so can be found at /usr/local/lib/libmesos.so, but if it is not found at the default location, you can find it at <MESOS_HOME>/build/src/.libs/libmesos.so.

2. Next, open `<SPARK_HOME>/conf/spark-defaults.conf` and define the variable, `spark.executor.uri=`. The value of this variable will be the location of the Spark binaries, either accessed via `http://` or `hdfs://` (Hadoop) or `s3://` (Amazon S3 at http://aws.amazon.com/s3/).

3. This variable is required for Mesos slave nodes. These nodes require Spark binaries for executing Spark jobs. Our Spark jobs will fail in case we do not provide the correct URL of the Spark binaries. Since `http` is the simplest one, we will directly reference the Spark binaries available at the Spark website. Define the following variable and its value in the `<SPARK_HOME>/conf/spark-defaults.conf` file:

   ```
   spark.executor.uri= http://d3kbcqa49mib13.cloudfront.net/spark-1.3.0-bin-hadoop2.4.tgz
   ```

   > In production systems, it is recommended to upload Spark binaries on HDFS, which should be in same network/subnet as of Mesos slave nodes.

4. Finally, we will execute our Spark job in the same manner as we did it before using `spark-submit`, where everything remains the same except the URL of the `--master` parameter needs to be changed to the master node of the Mesos cluster. For example, the samples from the *Your first Spark program* section of *Chapter 1, Installing and Configuring Spark and Spark Streaming*, can be submitted to Mesos cluster by executing the following command on our Linux console from the location where we have saved/exported our `Spark-Examples` project:

   ```
   $SPARK_HOME/bin/spark-submit --class chapter.one.ScalaFirstSparkExample --master mesos://<Host-Name>:5050 Spark-Examples.jar
   ```

5. As soon as we execute the preceding command, we will see the logs coming on the console, and at the same time, the **Framework** tab of the Mesos monitoring console will also tell us the status of our job that would be similar to the following screenshot:

```
sumit@localhost $ $SPARK_HOME/bin/spark-submit --class chapter.one.ScalaFirstSparkExample --master mesos://10.63.221.3:5050 Spark-Examples.jar
Spark assembly has been built with Hive, including Datanucleus jars on classpath
Creating Spark Configuration
Creating Spark Context
15/07/24 01:20:00 WARN NativeCodeLoader: Unable to load native-hadoop library for your platform... using builtin-java classes where applicable
I0724 01:20:01.701756  3787 sched.cpp:157] Version: 0.22.1
I0724 01:20:01.708050  3790 sched.cpp:254] New master detected at master@10.63.221.3:5050
I0724 01:20:01.708807  3790 sched.cpp:264] No credentials provided. Attempting to register without authentication
I0724 01:20:01.711602  3790 sched.cpp:448] Framework registered with 20150724-011300-64831242-5050-3364-0000
Loading the Dataset and will further process it
Number of Lines in the Dataset 98
sumit@localhost $ |
```

The preceding screenshot shows the output of our Spark job on the console. The following illustrations shows the Mesos UI with the status of our Spark job; the Spark Streaming jobs/applications can also be executed using the same process without any changes:

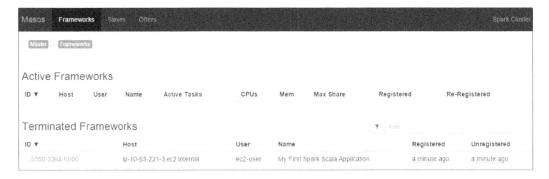

 In Spark 1.3, Spark driver can only be executed (using `spark-submit`) in the client mode, but in the recently released Spark 1.4.0 (June 2015), it can also be executed in the cluster mode. The output can be seen on Mesos UI at https://spark.apache.org/docs/1.4.0/running-on-mesos.html#cluster-mode.

Deploying on Hadoop or YARN

Hadoop 2.0 aka YARN was a complete change in the architecture. It was introduced as a generic cluster computing framework, entrusted with the responsibility for allocating and managing the resources required to execute the varied jobs or applications. It introduced new daemon services such as **resource manager** (**RM**), **node manager** (**NM**), and **application master** (**AM**), which are responsible for managing cluster resources, individual nodes, and respective applications.

 For more information on the architecture of YARN, please refer to http://hadoop.apache.org/docs/current/hadoop-yarn/hadoop-yarn-site/YARN.html.

YARN also introduced specific interfaces/guidelines for application developers where they can implement/follow and submit or execute their custom applications on the YARN cluster.

For more information on executing custom applications on YARN, refer to http://hadoop.apache.org/docs/current/hadoop-yarn/hadoop-yarn-site/WritingYarnApplications.html.

The Spark framework implements the interfaces exposed by YARN and provides the flexibility of executing the Spark applications on YARN. Spark applications can be executed in two different modes in YARN:

- **YARN client mode**: In this mode, the Spark driver executes the client machine (the machine used for submitting the job), and the YARN application master is just used for requesting the resources from YARN. The behavior of Spark application is the same as we have experienced in Mesos or standalone mode where all our logs and sysouts (println) are printed on the same console which is used for submitting the job.

- **YARN cluster mode**: In this mode, the Spark driver runs inside the YARN application master process, which is further managed by YARN on the cluster, and the client can go away just after submitting the application. Now as our Spark driver is executed on the YARN cluster, our application logs/sysouts (println) are also written in the log files maintained by YARN and not on the machine that is used to submit our Spark job.

Let's move forward and execute our Spark jobs on YARN. Perform the following steps for integrating and submitting Spark applications on YARN:

1. Download Hadoop 2.4 from https://archive.apache.org/dist/hadoop/common/hadoop-2.4.0/hadoop-2.4.0.tar.gz.

2. Extract TAR file to any of the directories and execute the following command on your Linux console:

   `export HADOOP_HOME = <path of the directory where Hadoop is extracted >`

   `export HADOOP_CONF_DIR = $HADOOP_HOME /etc/hadoop`

3. Next, we will set up Hadoop and YARN cluster. Hadoop/YARN can be set up in three different modes:
   - **Standalone mode**: To set up YARN in this mode, perform the steps provided at http://hadoop.apache.org/docs/r2.4.1/hadoop-project-dist/hadoop-common/SingleCluster.html#Standalone_Operation.

Deploying in Production

- **Pseudo-distributed mode**: For this mode, perform the steps provided at `http://hadoop.apache.org/docs/r2.4.1/hadoop-project-dist/hadoop-common/SingleCluster.html#Pseudo-Distributed_Operation`.
- **Fully distributed mode**: To set up YARN in this mode, perform the steps at `http://hadoop.apache.org/docs/r2.4.1/hadoop-project-dist/hadoop-common/SingleCluster.html#Fully-Distributed_Operation`.

Once our cluster is up and running in any of the specified modes, ensure that the following daemons/services are running in our Hadoop/YARN cluster:

- **Resource and node manager**: Open a new browser window and browse `http://<HOST-NAME/IP>:8088/cluster`. This should show the UI of the resource manager.
- **Name and data node**: Open a new browser window and browse `http://<HOST-NAME/IP>:50070/dfshealth.html`. This should show the UI of the name node that will provide options to browse and view Hadoop Distributed File System, data nodes, and various other Hadoop configurations.

4. Next, we will execute our first Spark Streaming example developed in the *Your first Spark Streaming program* section on YARN in client mode in *Chapter 2, Architecture and Components of Spark and Spark Streaming*. Follow the same steps as defined in the example, but for submitting our Spark Streaming job, use the following command where we have replaced `<SPARK-MASTER-URL>` with `yarn-client`:

```
$SPARK_HOME/bin/spark-submit --class chapter.two.ScalaFirstStreamingExample --master yarn-client Spark-Examples.jar
```

5. Although there is no change in the result or outcome of the preceding command, now we are leveraging YARN for our job execution and we can see the same in the ResourceManager UI by browsing `http://<HOST-NAME/IP>:8088/cluster`. This would be similar to the following screenshot:

Chapter 7

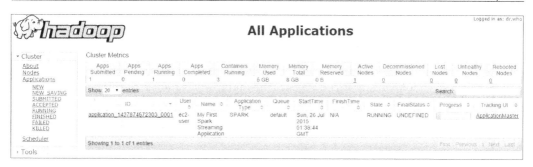

The preceding illustration shows the ResourceManager UI where we can see the detailed status, logs, and so on of all applications executed by YARN. You can click on **Tracking UI** or **ID** of the application for browsing the logs.

6. Next, for executing the same example (which we did in the previous step) in the YARN cluster mode, we need to just change `yarn-client` to `yarn-cluster` in our `spark-submit` command:

   `$SPARK_HOME/bin/spark-submit --class chapter.two.ScalaFirstStreamingExample --master yarn-cluster Spark-Examples.jar`

7. As `yarn-cluster` would execute our Spark drive in cluster mode, so all the logs printed on the console will be visible in the YARN logs and not on the console that we used to submit the job. You can browse the logs and see the results by clicking on the ID of the application on the ResourceManager UI and then on logs. The following screenshot shows the details of our Spark Streaming job on the YARN ResourceManager UI:

> The path of all the static or dynamic files that is processed by our Spark Streaming job should be available on HDFS. YARN will not read files from the local filesystem. For more information on deploying Spark and Spark Streaming application on YARN and Mesos, refer to https://spark.apache.org/docs/1.3.0/running-on-yarn.html and https://spark.apache.org/docs/1.3.0/running-on-mesos.html.

In this section, we discussed the steps involved in deploying our Spark Streaming applications in various cluster computing frameworks. Let's move toward the next section where we will discuss about the process and options available for deploying highly available and fault tolerant Spark Streaming applications.

High availability and fault tolerance

In this section, we will talk about the high availability and fault tolerance features of Spark and Spark Streaming in various kind of deployment models.

High availability in the standalone mode

In the standalone mode, Spark cluster, by default, is resilient to the failure of worker nodes/processes. As soon as the worker goes down, the master chooses another available worker and schedules the jobs, but what if the master itself goes down? Will it be the single point of failure? No!

Spark Standalone mode leverages ZooKeeper (http://zookeeper.apache.org/doc/trunk/zookeeperStarted.html) for leader election and provides the flexibility to create multiple/backup masters, which automatically takes up the role of master as soon as the current master goes down. In any case and at any point of time, there would be one and only one master serving the user requests.

Chapter 7

In order to enable the HA mode, configure SPARK_DAEMON_JAVA_OPTS in $SPARK_HOME/conf/spark-env.sh and add/append the following properties:

- spark.deploy.recoveryMode: The default value is NONE. Set it to ZOOKEEPER to enable the HA mode.
- spark.deploy.zookeeper.url: Provide the comma-separated IP:PORT of nodes where we have configured the ZooKeeper cluster. For example, 192.168.1.10:2181, 192.168.1.11:2181.
- spark.deploy.zookeeper.dir: The directory in ZooKeeper to store recovery state (the default value is spark).

> Ensure that we configure masters correctly with ZooKeeper instances; incorrect configurations may lead to undesirable results where masters may be unable to discover each other and every node will assume itself as master and start scheduling the jobs on worker nodes.

Once our master nodes are configured with ZooKeeper and our cluster is up and running, the client applications or worker process need not to worry about the current master node. They can specify the comma-separated list of all the configured master nodes in SparkContext so that our applications is registered with all the configured master nodes (spark://host:port,host1:port1…), and in case one of the master nodes goes down, the configuration remains valid then too.

Once the application registers itself with any of the active masters, the failovers are seamless and automatic; the application does not have to do anything.

The way it works is that the registration details of all applications is stored within the ZooKeeper and in case the current master goes down, the new master reads all the configurations including details about the registered applications from ZooKeeper and informs all the previously registered applications and workers about the change in leadership. So, the applications now do not have to worry about the current master nodes, and at the same time, it also provides the flexibility where we can bring up master nodes at any point of time without changing anything in the applications.

The following illustration shows the overall HA architecture process for the standalone deployment mode:

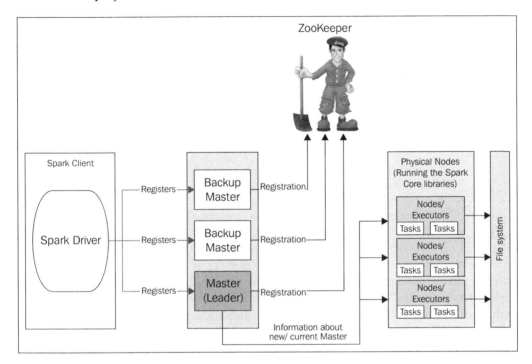

 For more information on HA in the standalone mode, refer to https://spark.apache.org/docs/1.3.0/spark-standalone.html.

High availability in Mesos or YARN

High availability in Apache Mesos and YARN works in similar manner as it does in the standalone mode. The overall architecture of achieving high availability also remains the same where Mesos/YARN also leverages ZooKeeper and provides the flexibility of having backup master/resource manager nodes, which can be automatically elected as leader by ZooKeeper in case of any failures with the current master node.

> For more information on HA in Apache Mesos and YARN, refer to http://mesos.apache.org/documentation/latest/high-availability/ and http://hadoop.apache.org/docs/current/hadoop-yarn/hadoop-yarn-site/ResourceManagerHA.html.

Fault tolerance

Fault tolerance is one of the critical architectures and design principles that enables the smooth functioning of any software system without any data loss, which may happen due to the malfunctioning of any component of the system.

As per Wikipedia (https://en.wikipedia.org/wiki/Fault_tolerance):

> *"Fault tolerance is the property that enables a system to continue operating properly in the event of the failure of (or one or more faults within) some of its components."*

There can be various types of faults within a system, for example, hardware failures, server crashes, software bugs, physical damages, or other flaws introduced to the system from outside/external sources. The objective of any fault tolerant system is to identify and design system in such a manner that it is capable enough to gracefully handle the identified or unidentified possibility of errors that may occur in future.

> For more information on fault tolerance in distributed systems, refer to https://en.wikipedia.org/wiki/Fault-tolerant_computer_system and http://www-itec.uni-klu.ac.at/~laszlo/courses/DistSys_BP/FaultTolerance.pdf.

Let's move forward and discuss the various intrinsic properties and features of Spark/Spark Streaming that provides fault tolerance.

Fault tolerance in Spark Streaming

Processing of streaming data in near real-time involves various components where we need to ensure that appropriate fault tolerance is already available or can be applied. Let's discuss the each of these components and their resiliency toward the various failures:

- **DStreams**: DStreams are nothing more than a wrapper around the continuous series of **RDDs (Resilient Distributed Datasets)**. RDDs themselves are immutable, deterministically recomputable distributed dataset. In event of loss of any partition of RDD due to the failure in worker nodes, the lost partition can be recomputed by applying the lineage of operations on the original dataset.

  All RDD transformations are deterministic, so irrespective of any failures in the cluster, the final RDD always remains the same. This is the default behavior of RDD that also enables DStreams to be resilient towards the failures.

- **Data receivers**: Unlike Spark batch processes that work on fault tolerant file systems such as S3 and HDFS, Spark Streaming receives most of the streaming data (except while reading from filesystem) over the network from near real-time streams where we have to consume and process the data at the same time.

  Once the data is successfully consumed and converted into DStreams there is no worry but while we consume and process the data, there is a possibility of failure happening at multiple levels. Spark provides the following guidelines for handling failures at the time of receiving or consuming data from network streams:

    - **Enabling replication**: The data received by the receivers can be replicated to the multiple spark executors on the worker nodes. We can enable replication by using appropriate `StorageLevels` (ending with _2 such as MEMORY_ONLY_2, MEMORY_AND_DISK_2, and more). Another benefit of replication is that the executors continue running the tasks on the RDD without waiting to recompute a lost partition.

- **Enabling write-ahead logs**: There could be instances where data is received but not replicated because it is still getting processed/transformed by the executors, and in case any failures occur at this time, there can be a complete loss of data.

  Spark provided write-ahead logs, where the input data received through receivers will be first saved to write-ahead logs and then only it will be processed by the executors. Write-ahead logs are nothing but the journal which is recommended to be maintained on the distributed and reliable file system like HDFS or S3. This also facilitates the recovery of data even after driver failures. It can be enabled by configuring `spark.streaming.receiver.writeAheadLog.enable=true` in our `SparkConf`. Needless to say that we have to pay some price, that is performance, for using this feature, but again it is the trade-off between the various non-functional requirements (performance versus fault tolerance).

  Disable replication while we enable write-ahead logs as replication will be provided by the underlying filesystem anyways.

- **Reliable receivers**: We should also think about using the sources and receivers that supports acknowledgments (known as **acking**) so that the data is reliably received and written to write-ahead logs such as Kafka, Flume, and others. For more info on receiver's reliability, refer to https://spark.apache.org/docs/1.3.0/streaming-custom-receivers.html.

For more information on fault tolerance in Spark Streaming, refer to https://spark.apache.org/docs/1.4.0/streaming-programming-guide.html#fault-tolerance-semantics and https://databricks.com/blog/2015/01/15/improved-driver-fault-tolerance-and-zero-data-loss-in-spark-streaming.html

- **Spark driver**: The Spark driver is another single point of failure. In the scenarios where the system running the Spark driver crashes, our whole job is also halted. This is a serious concern but it can be addressed by implementing/enabling the following guidelines/features:

    - **Deploy modes**: The spark-submit script is the most common script for submitting the jobs/application and launch SparkDriver. By default, SparkDriver runs/executes on the same machine which is used to launch the spark-submit script. Launching scripts on client machines is unsafe and does not survive failures. The spark-submit script provides an option to specify --deploy-mode where we can specify to deploy our driver on the worker nodes (cluster) or on external machine in the client mode (default mode). For example, we can execute our streaming job developed in *Chapter 2, Architecture and Components of Spark and Spark Streaming*, by specifying the following command on the standalone Spark cluster:

        ```
        $SPARK_HOME/bin/spark-submit --class chapter.two.
        ScalaFirstStreamingExample --master <MASTER-IP> --deploy-
        mode cluster Spark-Examples.jar
        ```

    - For YARN, we can just specify yarn-cluster against --master and jobs are launched in the YARN cluster:

        ```
        $SPARK_HOME/bin/spark-submit --class chapter.two.
        ScalaFirstStreamingExample --master yarn-cluster Spark-
        Examples.jar
        ```

    - In the cluster mode, our driver is launched from one of the worker nodes in the cluster itself that makes it much reliable and can definitely survive failures as cluster itself is highly available and resilient to failures.

> In Mesos, the launching driver in cluster mode is only available in Spark 1.4.x.

- **Automatic restart**: In cluster mode, we can also add one more parameter, `--supervise`, to our `spark-submit` script, which will automatically restart our driver in case it exits with a non-zero code or may be failure of the node executing the driver. For example, we can execute our streaming jobs developed in *Chapter 2, Architecture and Components of Spark and Spark Streaming*, by specifying the following command on the standalone Spark cluster:

   ```
   $SPARK_HOME/bin/spark-submit --class chapter.two.
   ScalaFirstStreamingExample --master <MASTER-IP> --deploy-mode
   cluster --supervise Spark-Examples.jar
   ```

For YARN, refer to `http://hadoop.apache.org/docs/r2.4.1/hadoop-yarn/hadoop-yarn-site/ResourceManagerRestart.html`, and for Mesos, we can use `https://github.com/mesosphere/marathon` for automatic restart of jobs.

In this section, we have discussed about the various features provided by Spark for ensuring that our real-time jobs are resilient to failures and can be executed 24 x 7 without minimum or no manual intervention. Let's move forward and understand the security and monitoring aspects of the Spark Streaming applications that is another important aspect of enterprise deployments.

Monitoring streaming jobs

Spark provides different ways to monitor our Spark Streaming job. Though monitoring also depends upon the underlying cluster manager, at the same time, Spark itself produces a lot of information about our jobs that can be captured, analyzed, and further help to monitor and tune our jobs:

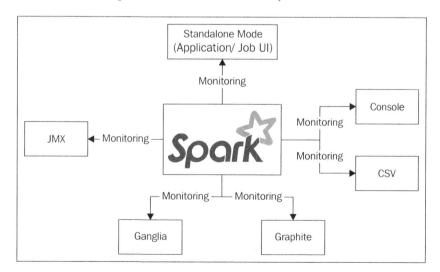

Deploying in Production

The preceding illustration shows the various monitoring options available with Spark, by default. At high level, the monitoring options can be categorized into two different categories, either of which can be leveraged to monitor our streaming jobs deployed in the standalone mode.

Application or job UI

Spark hosts a Web-UI over HTTP (by default, on 8080) on the master node, which shows the status and history of all the active or completed jobs and the list of worker nodes available/ready for accepting the jobs.

The preceding screenshot shows the Web-UI hosted on the master node and the details about workers and applications.

Chapter 7

We can click on **Worker id** of any of the **Workers** section to see the list of in-process and completed executors by that specific worker.

The preceding screenshot shows the details presented by the Web-UI of the specific node/worker.

Each SparkContext also launches a separate WEB-UI, which provides useful details and statistics about the jobs running on the Spark cluster binded to that specific SparkContext. By default, this UI is available at `http://<IP-MasterNode>:4040/` and can be accessed by clicking on any of the jobs listed in the **Running Applications** tab on the Web-UI hosted by the master node.

Deploying in Production

In case there are multiple SparkContext running/active at a time on the Spark hosts, they are bind to the successive ports (4041/4042/ and so on).

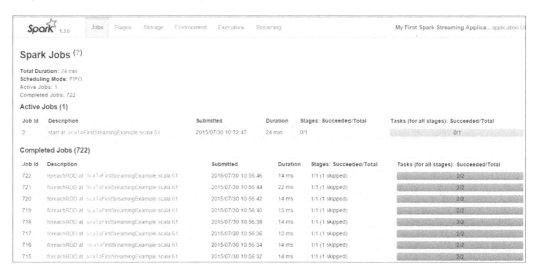

The preceding screenshot shows the WEB-UI presented by each of the active SparkContext within the cluster. Basically, it divulges the following information about each and every job which is currently running on the cluster:

- **Jobs**: This gives the list of active and completed Spark jobs.
- **Stages**: This shows the list of scheduler stages and tasks being executed by the executors.
- **Storage**: This gives a summary of RDD sizes (in memory and disk), memory usage, and cached partitions.
- **Environment**: This shows various configured runtime configurations and properties.
- **Executors**: This gives information about the executors running on that machine and its further details such as the memory used and completed tasks. We can also browse logs and at the same time take thread dumps for analyzing any deadlocks.
- **Streaming**: This shows the statistics, summary, and performance details from the last 100 batches executed by that specific Spark Streaming context. It provides details about the receiver and batch processing statistics such as processing time and scheduling delays for the last 100 batches.

All this information is only available while the SparkContext/application is active. As soon as SparkContext is killed, all this information is lost. In case we need to still see these details even after the job is completed, we need to enable the event logging and configure the following parameters in our `$SPARK_HOME/conf/spark-default.conf` file:

- The `spark.eventLog.enabled=true` parameter enables the event logging of applications jobs
- The `spark.eventLog.dir=hdfs://localhost:9000/jobHistory` parameter defines the location of these logs on HDFS

Once we have enabled the event logging, we can bring up our history server by configuring the following parameter in our `Spark-default.conf` file:

- `spark.history.fs.logDirectory=hdfs://localhost:9000/jobHistory`: This parameter defines the location of the directory where event logs are stored

Finally, we will bring up our history server by executing the following command on our Linux console:

`$SPARK_HOME/sbin/start-history-server.sh`

Our history server, by default, is hosted on port 18080 and provides similar details to what we have seen in Web-UI for the running/live SparkContext, with the only difference that it shows only completed or incompleted jobs. Incompleted jobs are those that are forcefully halted by the admin and have not been gracefully shut down by invoking `stop()` on `SparkContext` or `StreamingContext`.

The preceding screenshot shows the UI of the **History Server**. We can click on any of the completed or incompleted jobs and can analyze the same details as we have seen for live/in-process jobs being executed by our Spark cluster.

Integration with other monitoring tools

Spark, by default, also provides integration with other monitoring tools such as Ganglia and Graphite. It provides a flexible, extendable, and configurable metrics system based on the Coda Hale metrics library (`https://github.com/dropwizard/metrics`).

Let's follow these steps to enable JMX visualization for viewing the statistics exposed by the Spark executors:

1. Rename `$SPARK_HOME/conf/metrics.properties.template` to `$SPARK_HOME/conf/metrics.properties`.

2. Open and edit `$SPARK_HOME/conf/metrics.properties` and uncomment the following properties:

   ```
   *.sink.jmx.class=org.apache.spark.metrics.sink.JmxSink
   master.source.jvm.class=org.apache.spark.metrics.source.JvmSource
   worker.source.jvm.class=org.apache.spark.metrics.source.JvmSource
   driver.source.jvm.class=org.apache.spark.metrics.source.JvmSource
   executor.source.jvm.class=org.apache.spark.metrics.source.JvmSource
   ```

3. Open and edit the `$SPARK_HOME/conf/spark-default.conf` file and add the following properties for enabling the remote connection to JMX:

   ```
   spark.executor.extraJavaOptions=-Dcom.sun.management.jmxremote.port=9999 -Dcom.sun.management.jmxremote.authenticate=false -Dcom.sun.management.jmxremote.ssl=false -Djava.rmi.server.hostname=<Host-Name>
   ```

4. We are done! Now execute your streaming jobs and open any JMX client such as JConsole, VisualVM, and others and use port <Host-Name>:9999 to connect and see various statistics about your streaming job. The following screenshot has the JConsole UI that shows the various metrics/statistics exposed by Spark JMX sink:

 For more information about monitoring in Spark, refer to https://spark.apache.org/docs/1.3.0/monitoring.html.

Summary

In this chapter, we discussed in detail about various deployment models supported by Spark and Spark Streaming. We also discussed various other critical and important aspects of production deployments such as high availability, fault tolerance, and monitoring.

Index

A

accumulators
  about 52
  URL 52
acking 163
actions 60
administrators/developers
  tools 18
  utilities 18
algorithms, for classification
  references 55
Amazon Kinesis
  URL 54
Amazon S3
  URL 153
Apache access logs
  URL 80
Apache Cassandra
  configuring 110, 111
  installing 110, 111
  integration with 110
  URL 110
Apache Cassandra 2.1.7
  URL 110
Apache Giraph
  URL 136
Apache Hadoop
  URL 28
Apache Maven
  URL 150

Apache Mesos
  configuring 150-152
  deploying on 149
  installing 150-152
  Spark applications, executing on 152-154
  Spark applications, integrating on 152-154
  URL 1
Apache Spark
  URL 1
application master (AM) 154
architecture, YARN
  URL 154
assembly lines 27

B

batch jobs
  examples 25
  references 26
batch processing
  about 8
  defining 25, 26
  URL 23
  versus real-time data processing 24
batch processing systems
  complexity 26
  Distributed processing 26
  Enterprise constraints 26
  Fault tolerant 26
  Large data 26
  Scalability 26
business intelligence (BI)
  about 27
  URL 27

C

Cassandra Core driver
  URL 112
classification models
  URL 55
clustering algorithms
  URL 55
cluster management
  $SPARK_HOME/conf 19
  $SPARK_HOME/sbin 19
  about 19
collaborative filtering
  URL 55
components, Flume
  Channel 67
  Interceptor 68
  Sink 68
  Source 67
components, Spark
  RDD 12
  SparkContext 12
components, Spark cluster
  Cluster manager/Spark master 8
  Driver 8
  Spark worker 8
components, Spark Streaming
  Batch 34
  data receivers 162
  DStreams 162
  Input data streams 33, 34
  Output data streams 34
  Spark Core engine 34
  Spark Streaming 34
configuration parameters
  URL 96
CQL (Cassandra Query Language) 111
CQLSH 111
Curator
  URL 158
custom applications, on YARN
  URL 155

custom sink
  URL 68

D

data analysis 79
DataFrames
  about 123
  URL 123
data lineage 59
data, loading
  about 65, 66
  Flume architecture 67-69
DataStax
  about 110
  URL 110
deployment models, Spark
  Apache Mesos 10
  Hadoop YARN 10
  Standalone mode 9
Directed Acyclic Graph (DAG)
  about 53
  URL 53
discretized streams 58, 63, 64
distributed log file processing
  architecture, defining 77
DStream
  higher-order functions 82
  operations 82
DStream API
  URL 64

E

Eclipse Luna (4.4)
  URL 7
edges 136
Elasticsearch
  URL 129
ETL (Extraction, Transformation,
    and Loading) 99

F

fastutil
  URL 95
fault tolerance
  about 158-161
  in Spark Streaming 162-165
  URL 161
fault tolerance, in Spark Streaming
  URL 163
Flume
  configuring 69-72
  installing 69-72
  references 67, 72, 73
Flume events
  consuming, via Spark configuration 73-76
Flume interceptors
  URL 68
Flume sinks
  URL 68
fully distributed mode, Hadoop/YARN
  URL 156
Fume
  URL 54

G

GraphLab
  URL 136
graph-parallel computations 47
Graphs
  URL 136
GraphX API
  defining 137-139
guidelines, Spark
  automatic restart 165
  reliable receivers 163
  replication, enabling 162
  Spark driver 164
  write-ahead logs, enabling 163

H

hacking
  URL 65
Hadoop
  about 23
  deploying on 154-158
  using 99
Hadoop 2.4
  URL 155
Hadoop 2.4.0 distribution
  references 104
  URL 103
Hadoop and HDFS
  URL 105
Hadoop YARN
  URL 10
HA, in Apache Mesos and YARN
  URL 161
HA, in standalone mode
  URL 160
HA mode
  enabling 159
hardware requirements, Spark
  CPU 3
  disk 4
  network 4
  operating system 4
  RAM 3
HDFS (Hadoop Distributed File System) 100
high availability
  about 158
  in Mesos 160
  in standalone mode 158, 159
  in YARN 160
high-order functions
  URL 88

J

Java APIs
  defining, in SparkConf 49
  defining, in SparkContext 49
Java-specific RDD classes
  URL 51
JSON editor
  URL 124
Just Bunch of Disks (JBOD) 4

K

Kafka
  URL 54
key performance indicators (KPI)
  URL 25
Kryo Serialization
  URL 94
Kyro documentation
  URL 94

L

latency 27
libraries
  URL 112

M

MapReduce programming model
  URL 23
Mesos
  URL 152
Mesos UI
  URL 154
micro-batching 90
modes, Hadoop/YARN
  fully distributed mode 156
  pseudo-distributed mode 156
  standalone mode 155
MongoDB
  URL 129
monitoring, in Spark
  URL 171
MQTT
  URL 54

N

Naive Bayes
  URL 55
node manager (NM) 154
nodes 136
non-RAID architectures
  URL 4

O

Operational intelligence (OI)
  about 27
  URL 27
operations, DStreams
  incremental aggregation/stateful
    processing 64
  output operations 64
  transformations 64
  windowing 64
optimization techniques
  URL 95
Oracle Java 7
  URL 6

P

PageRank
  URL 57
partitioning and parallelism
  URL 94
performance features
  URL 96
performance tuning
  about 93
  parallelism 93
  partitioning 93
  serialization 94
  Spark memory tuning 95
Point of Sale (POS) systems 27
Pregel, Google
  URL 136
property graph model 136
pseudo-distributed mode, Hadoop/YARN
  URL 156
Python APIs
  URL 46

R

Rapid Application Development (RAD) 1
RDD
  about 58
  action functions 60
  actions 60
  defining 58
  fault tolerance 59
  features 59
  functions 60
  persistence 61
  references 61
  shuffling 62
  storage 61
  transformations 60
RDD, Java APIs
  defining 50
RDD, Scala APIs
  defining 49, 50
  URL 49
real-time data processing
  defining 26-28
  versus batch processing 24
real-time data processing systems
  limitations 27
real-time systems
  examples 27
receiver's reliability
  URL 163
reduce function 92
resilient distributed datasets. *See* RDD
resource manager (RM) 154

S

Scala
  URL 3
Scala 2.10.5 compressed tarball
  URL 6
Scala APIs
  defining, in SparkConf 48
  defining, in SparkContext 48
Scala code
  URL 7

schema, using reflection
  URL 127
Seq
  URL 114
serialization 94
shortest path
  URL 57
shuffling
  about 62
  operations 62
  URL 62
slave nodes
  references 19
software development life cycle (SDLC) 147
software requirements, Spark
  defining 5
  Eclipse, installing 7
  Java, installing 6
  Scala, installing 6, 7
  Spark, installing 5
Spark
  about 1
  architecture 28
  client APIs 46, 47
  configuring, for integration
    with Cassandra 112
  configuring, to consume
    Flume events 73-76
  Data storage layer 30
  extensions 54
  hardware requirements 2
  installing 2
  layered architecture 30, 31
  libraries 54
  packaging structure 46, 47
  references 21, 52
  Resource manager APIs 30
  software requirements 5
  Spark Core 46
  Spark Core libraries 31
  Spark libraries/extensions 46, 47
  URL 148
  using 99
  versus Hadoop 29, 30

[177]

Spark 1.2
  URL 56
Spark and Flume integration
  URL 77
Spark applications
  YARN client mode 155
  YARN cluster mode 155
Spark binaries
  using 10
Spark-Cassandra connector
  URL 112
Spark-Cassandra Java library
  URL 112
Spark cluster
  configuring 8-11
  running 8-11
  setting up, URL 22
Spark Community
  URL 96
Spark compressed tarball
  URL 5
SparkConf
  URL 48
SparkContext
  about 12
  URL 12
Spark Core
  defining 48
  packages 51-53
Spark Core packages
  URL 51
Spark deployment models
  defining 148, 149
  deploying, on Apache Mesos 149
  deploying, on Hadoop 154-158
  deploying, on YARN 154-158
Spark extensions
  installing 8
Spark GraphX
  about 57
  integrating, with Spark Streaming 140-146
  packages 57
  packages, references 57
  URL 47
  used, for graph analysis 135-137

Spark jobs
  coding, for streaming web logs
    in Cassandra 113-119
  submitting 20
Spark libraries/extensions
  Spark GraphX 47
  Spark MLlib 47
  Spark SQL 47
  Spark Streaming 46
Spark master and worker
  URL 20
Spark memory tuning
  about 95
  executor memory 96
  garbage collection 95
  object sizes 95
  RDDs, caching 96
Spark MLlib
  about 55
  URL 47
Spark program
  defining 12
  Spark jobs, coding in Java 16-18
  Spark jobs, coding in Scala 12-16
Spark SQL
  about 56, 123-128
  features 123
  integrating, with streams 129-135
  URL 56
Spark SQL packages
  references 56, 57
Spark SQL, with Cassandra
  URL 129
Spark Streaming
  about 8, 45, 54
  architecture 32
  defining 32
  functions 54
  high-level architecture 32-34
  output operations 100-109
  sub packages, defining 54
  URL 32
Spark Streaming APIs
  URL 55

Spark Streaming job
  deploying 76, 77
  packaging 76, 77
Spark Streaming program
  client application 40, 41
  defining 35
  Spark Streaming job, deploying 42, 43
  Spark Streaming job, packaging 42, 43
  Spark Streaming jobs, coding in Java 38-40
  Spark Streaming jobs, coding in Scala 35-37
SPOF (single point of failure) 99
spoofing
  URL 65
SQLContext
  URL 126
standalone mode 10
standalone mode, Hadoop/YARN
  URL 155
StorageLevel class
  URL 61
streaming data
  querying, in real time 122
streaming jobs
  application 166-169
  integration, with other
      monitoring tool 170, 171
  job UI 166-169
  monitoring 165, 166
support vector machines
  URL 55
SVD
  URL 57

T

Tachyon
  URL 96
The Open Group Architecture
      Framework (TOGAF) 121
transformation functions
  applying 80
  defining 80
  functional operations 82-88
  log streaming, simulating 80-82
  transform operations 89, 90
  windowing operations 90-92

troubleshooting
  about 20
  classpath issues 21
  common exceptions 21
  PORT numbers, configuring 21

U

utility JAR file
  URL 131

V

vertices 136

Y

YARN
  deploying on 154-158
  URL 165
YARN and Mesos
  references 158

Z

Zero MQ
  URL 55
ZooKeeper
  URL 158

Thank you for buying
Learning Real-time Processing with Spark Streaming

About Packt Publishing

Packt, pronounced 'packed', published its first book, *Mastering phpMyAdmin for Effective MySQL Management*, in April 2004, and subsequently continued to specialize in publishing highly focused books on specific technologies and solutions.

Our books and publications share the experiences of your fellow IT professionals in adapting and customizing today's systems, applications, and frameworks. Our solution-based books give you the knowledge and power to customize the software and technologies you're using to get the job done. Packt books are more specific and less general than the IT books you have seen in the past. Our unique business model allows us to bring you more focused information, giving you more of what you need to know, and less of what you don't.

Packt is a modern yet unique publishing company that focuses on producing quality, cutting-edge books for communities of developers, administrators, and newbies alike. For more information, please visit our website at www.packtpub.com.

About Packt Open Source

In 2010, Packt launched two new brands, Packt Open Source and Packt Enterprise, in order to continue its focus on specialization. This book is part of the Packt Open Source brand, home to books published on software built around open source licenses, and offering information to anybody from advanced developers to budding web designers. The Open Source brand also runs Packt's Open Source Royalty Scheme, by which Packt gives a royalty to each open source project about whose software a book is sold.

Writing for Packt

We welcome all inquiries from people who are interested in authoring. Book proposals should be sent to author@packtpub.com. If your book idea is still at an early stage and you would like to discuss it first before writing a formal book proposal, then please contact us; one of our commissioning editors will get in touch with you.

We're not just looking for published authors; if you have strong technical skills but no writing experience, our experienced editors can help you develop a writing career, or simply get some additional reward for your expertise.

Fast Data Processing with Spark

ISBN: 978-1-78216-706-8   Paperback: 120 pages

High-speed distributed computing made easy with Spark

1. Implement Spark's interactive shell to prototype distributed applications.

2. Deploy Spark jobs to various clusters such as Mesos, EC2, Chef, YARN, EMR, and so on.

3. Use Shark's SQL query-like syntax with Spark.

Machine Learning with Spark

ISBN: 978-1-78328-851-9   Paperback: 338 pages

Create scalable machine learning applications to power a modern data-driven business using Spark

1. A practical tutorial with real-world use cases allowing you to develop your own machine learning systems with Spark.

2. Combine various techniques and models into an intelligent machine learning system.

3. Use Spark's powerful tools to load, analyze, clean, and transform your data.

Please check **www.PacktPub.com** for information on our titles

Fast Data Processing with Spark
Second Edition

ISBN: 978-1-78439-257-4   Paperback: 184 pages

Perform real-time analytics using Spark in a fast, distributed, and scalable way

1. Develop a machine learning system with Spark's MLlib and scalable algorithms.
2. Deploy Spark jobs to various clusters such as Mesos, EC2, Chef, YARN, EMR, and so on.
3. This is a step-by-step tutorial that unleashes the power of Spark and its latest features.

Getting Started with Apache Maven

ISBN: 978-1-78216-572-9   Duration: 02:15 minutes

Design and manage simple to complex Java projects effectively using Apache Maven's project object model

1. Covers everything from basic dependencies to complex multi-module projects.
2. Demonstrates the key concept of project building logically.
3. Loaded with examples, motivated by typical build challenges.

Please check **www.PacktPub.com** for information on our titles

Made in the USA
Columbia, SC
11 August 2018